SpringerBriefs in Math

SpringerBriefs in Mathematics showcases expositions in all areas of mathematics and applied mathematics. Manuscripts presenting new results or a single new result in a classical field, new field, or an emerging topic, applications, or bridges between new results and already published works, are encouraged. The series is intended for mathematicians and applied mathematicians.

For further volumes:
http://www.springer.com/series/10030

Qi He • Le Yi Wang • G. George Yin

System Identification Using Regular and Quantized Observations

Applications of Large Deviations Principles

Qi He
Department of Mathematics
University of California
Irvine, California, USA

Le Yi Wang
Department of Electrical
and Computer Eng
Wayne State University
Detroit, Michigan, USA

G. George Yin
Department of Mathematics
Wayne State University
Detroit, Michigan, USA

ISSN 2191-8198
ISSN 2191-8201 (electronic)
ISBN 978-1-4614-6291-0
ISBN 978-1-4614-6292-7 (eBook)
DOI 10.1007/978-1-4614-6292-7
Springer New York Heidelberg Dordrecht London

Library of Congress Control Number: 2012955366

Mathematics Subject Classification (2010): 93E12, 93E10, 93E03, 60F10

Printed on acid-free paper

Springer is part of Springer Science+Business Media (www.springer.com)

To my parents, Guozhen Cao and Qiumin He

Qi He

To my wife, Hong, and daughters, Jackie and Abby, for always believing, without evidence, that what I am doing might be interesting and useful

Le Yi Wang

To Meimei for her support and understanding

George Yin

Preface

This monograph develops large deviations estimates for system identification. It treats observations of systems with regular sensors and quantized sensors. Traditional system identification, taking noisy measurements or observations into consideration, concentrates on convergence and convergence rates of estimates in suitable senses (such as in mean square, in distribution, or with probability one). Such asymptotic analysis is inadequate in applications that require a probabilistic characterization of identification errors. This is especially true for system diagnosis and prognosis, and their related complexity analysis, in which it is essential to understand probabilities of identification errors. Although such probability bounds can be derived from standard inequalities such as Chebyshev and Markov inequalities, they are of polynomial type and conservative. The large deviations principle provides an asymptotically accurate characterization of identification errors in an exponential form and is appealing in many applications.

The large deviations principle is a well-studied subject with a vast literature. There are many important treatises on this subject in probability, statistics, and applications. This brief monograph adds new contributions from the point of view of system identification, with emphasis on the features that are strongly motivated by applications. Three chapters are devoted to case studies on battery diagnosis, signal processing for medical diagnosis of patients with lung or heart diseases, and speed estimation for permanent magnet direct current (PMDC) motors with binary-valued sensors. The basic methods and estimates obtained can of course be applied to a much wider range of applications.

This book has been written for researchers and practitioners working in system identification, control theory, signal processing, and applied probability. Some of its contents may also be used in a special-topics course for graduate students.

Having worked on system identification over a time horizon of more than twenty years, we thank many of our colleagues who have worked with us on related problems and topics. During these years, our work has been supported by (not at the same time) the National Science Foundation, the Department of Energy, the Air Force Office of Scientific Research, the Army Research Office, the Michigan Economic Development Council, and many industry gifts and contracts. Their support and encouragement are gratefully acknowledged. We would like to thank the reviewers for providing us with insightful comments and suggestions. Our thanks also go to Donna Chernyk for her great effort and expert assistance in bringing this volume into being. Finally, we thank the many Springer professionals who assisted in many logistic steps to bring this book to its final form.

Contents

Notation and Abbreviations

$\lvert A\rvert$ and $\lvert x\rvert$	Euclidean norms
$\overline{B}$	Closure of set B
B°	Interior of set B
$C_0([0,T];\mathbb{R}^n)$	Space of continuous functions
$C^2(\mathbb{R}^n)$	Class of real-valued C^2 function on $\mathbb{R}^n$
$D_N(t_0)$	$= (d(t_0),\ldots,d(t_0+N-1))'$
$H(\cdot)$	H functional
$H(v\vert u)$	Relative entropy of probability vector v with respect to u
$I(\cdot)$	Rate function
$I^b(\cdot)$	Rate function for binary sensor
$I^q(\cdot)$	Rate function for quantized sensor
$I^r(\cdot)$	Rate function for regular sensor
K	Positive constant
LDP	Large deviations principle
$Y_N(t_0)$	$= (y(t_0),\ldots,y(t_0+N-1))'$
$d(\cdot,\cdot)$	Metric in a Euclidean space or a function space
$d(t)$	Noise
∇h	Gradient of h
$u(t)$	Input sequence
v'	Transpose of $v \in \mathbb{R}^{l_1\times l_2}$ with l_1, $l_2 \geq 1$
$y(t)$	Output or observation
i.i.d.	Independent and identically distributed
w.p.1.	With probability one

Φ_0	$= \Phi_{m_0}(t_0)$
$\Phi_N(t_0)$	$= (\varphi(t_0), \ldots, \varphi(t_0 + N - 1))'$
$\widetilde{\Phi}_N(t_0)$	$= (\widetilde{\varphi}(t_0), \ldots, \widetilde{\varphi}(t_0 + N - 1))'$
χ_A	Indicator function of the set A
θ	Modeled part of the parameter
$\widetilde{\theta}$	Unmodeled dynamics
$\widehat{\theta}_n$	Parameter estimate
$\varphi(t)$	$= (u(t), \ldots, u(t - m_0 + 1))'$
$\widetilde{\varphi}(t)$	$= (u(t - m_0), u(t - m_0 - 1), \ldots, \ldots,)'$
$\langle \cdot, \cdot \rangle$	Inner product
$\mathbf{1}$	$(1, \ldots, 1)'$
$\square$	End of proof
$\lVert \cdot \rVert_1$	$\lVert a \rVert = \sum_{i=1}^{\infty} \lvert a_i \rvert$
$\lVert \cdot \rVert_\infty$	Supremum norm on function space $C_0([0, T]; \mathbb{R}^n)$
$\lVert \cdot \rVert_{H^\infty}$	H^∞ norm

1
Introduction and Overview

Traditional system identification taking noise measurement into consideration concentrates on convergence in suitable senses (such as in mean square, in distribution, or with probability one) and rates of convergence. Such asymptotic analysis is inadequate in applications that require precise probability error bounds beyond what are provided by the law of large numbers or the central limit theorem. Especially, for system diagnosis and prognosis and their related complexity analysis, it is essential to understand probabilities of identification errors over a finite data window. For example, in real-time diagnosis, parameter values must be evaluated to determine whether they belong to a "normal" region or a "fault" has occurred. This set-based identification amounts to hypothesis testing, which relies on an accurate probabilistic characterization of parameter estimates.

On the other hand, worst-case system identification treats identification errors in a set-membership framework. Within a finite data window, it characterizes identification accuracy by specifying the uncertainty set that contains the parameters. The concept of rare occurrence can also be accommodated. In both stochastic and worst-case frameworks, a rigorous characterization of small-probability events or rare occurrence is of practical importance in guiding resource allocation and reliability assessment. The large deviations principles (LDPs) are a proper tool to provide such a characterization.

Q. He et al., *System Identification Using Regular and Quantized Observations*, SpringerBriefs in Mathematics, DOI 10.1007/978-1-4614-6292-7_1,

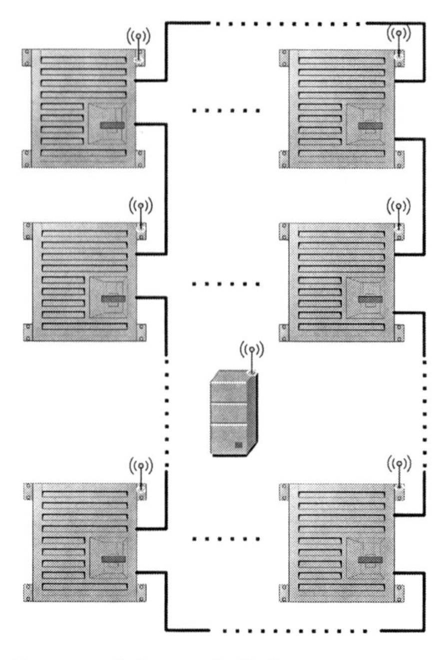

FIGURE 1.1. Battery modules and their measurements through wireless communications

Example 1.1 Management of battery systems plays a pivotal role in electric and hybrid vehicles and in support of distributed renewable energy generation and smart grids. A battery system contains many battery cells or modules. In large-scale battery packages, hundreds or even thousands of battery cells are interconnected with wired or wireless measurements of current/voltage/temperature of cells and modules (Fig. 1.1). The state of charge (SOC) and state of health (SOH), together with internal impedance (R), polarization constant (K), and open-circuit voltage (E_0), are essential parameters for characterizing the conditions of the battery system and guiding control and management strategies. For example, the SOH of a battery cell is typically evaluated by its maximum capacity Q. When Q is reduced below a threshold of its rated value, say 75%, the battery will be retired from the vehicle. This decision requires estimation of Q during routine operations, together with its error characterization in a probabilistic form under a finite data window, so that reliability of the decision can be assessed. Recently, there have been substantial efforts in developing new methods for real-time battery characterization using system identification methods [42, 46]. However, accurate probabilistic characterization of estimation errors remains an open issue in battery management systems (BMS).

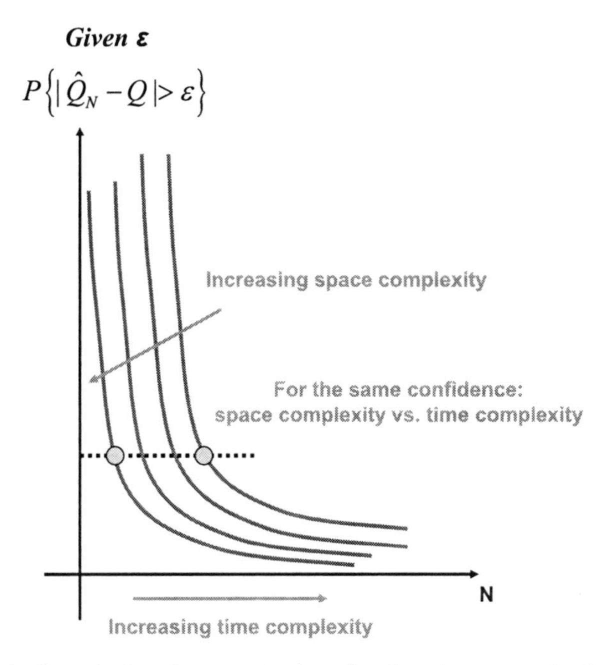

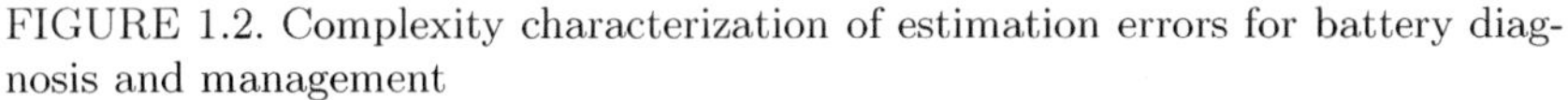

FIGURE 1.2. Complexity characterization of estimation errors for battery diagnosis and management

In consolidating issues of measurement costs and diagnostic accuracy, it is desirable to establish estimation properties in a complexity framework. The number N of samples in a given time interval represents the time complexity of the data flow. Then the measurement quantization resolution is a measure of the space complexity. Together, they define the data flow rates in bits per second (bps), which in turn affect communication bandwidth consumption, data storage requirements, computational speed demands, I/O interfacing costs, and data shuffling overhead.

On the other hand, the accuracy and reliability of decisions are related to quantization resolution and sampling rates, as illustrated in Fig. 1.2. Establishing this relationship is the essential task of this monograph. Such a characterization can be used to decide the benefits of increasing quantization accuracy and sampling rates, both of which will increase data flow rates, and to support hardware selection and design tradeoffs.

This book investigates identification errors under a probabilistic framework. By considering both space complexity in terms of signal quantization and time complexity with respect to data window sizes, our study provides a new perspective in understanding the fundamental relationship between probabilistic errors and resources, which may represent data sizes in computer use, computational complexity in algorithms, sample sizes in statistical analysis, channel bandwidths in communications, etc. This relationship is derived by establishing the LDP for both regular and quantized identification that links binary-valued data at one end for the lowest space complexity and regular sensors at the other. Under some mild conditions, we obtain large deviations upper and lower bounds. Our results accommodate independent and identically distributed (i.i.d.) noises, as well as more general classes of mixing-type noise sequences. In this book, we first treat i.i.d. noises, which enables us to give a clear presentation of the large deviations rate functions. Then further results on correlated noises are dealt with.

Traditional system identification under regular sensors is a mature field with many significant results concerning identification errors, convergence, convergence rates, and applications. For related results and literature, we refer the reader to the books by Aström and Wittenmark [1], Caines [8], Chen and Guo [10], Kumar and Varaiya [28], Ljung and Söderström [37], and Solo and Kong [51], among others; see also the most recent work of Ljung, Hjalmarsson, and Ohlsson [36]. For a general stochastic approximation framework for analyzing recursive identification algorithms, the reader may consult [31] and the references therein. System identification under quantized observations is a relatively new field that began to draw substantial attention only in the past decade and has achieved significant progress recently; see [58–61] and references therein. A recent monograph contains coverage of these results [62].

This book departs from the existing literature on system identification from several perspectives. First, rather than concentrating on convergence of parameter estimates and treating asymptotic distributions of centered and suitably scaled sequences of the estimates, we investigate large deviations of the parameter estimators, which provide an accurate characterization of the dependence of probabilistic identification errors on data window sizes. Second, our study employs different techniques from traditional tools for identification convergence analysis. Third, this book deals with the LDP on quantized identification, which represents a first attempt in this direction. Finally, our results target different applications and delineate the rate functions of the estimators that resolve some intriguing questions such as complexities of sensor types and relations to the Cramér–Rao bounds.

To elaborate on our motivations in a generic setting, consider the following scenario. Suppose a sequence of vector-valued estimates $\{\widehat{\theta}_n\}$ of the true parameter θ has been generated by an identification algorithm. Under suitable (and rather broad) conditions, the sequence is strongly consistent in the sense of convergence with probability one (w.p.1). Such estimates are known as consistent (convergence to the true value). We may further establish that $\sqrt{n}(\widehat{\theta}_n - \theta)$ converges in distribution to a normal random variable with mean zero and a derived asymptotic covariance. The scaling factor together with asymptotic covariances is a measure of convergence rates. This measure, however, cannot specify probability errors in terms of data window size. Suppose that we are interested in the quality of estimation in terms of

$$P(|\widehat{\theta}_n - \theta| > \varepsilon)$$

for a given $\varepsilon > 0$. Strong consistency or the asymptotic normality can affirm

$$P(|\widehat{\theta}_n - \theta| > \varepsilon) \to 0 \quad \text{as} \quad n \to \infty$$

without more precise bounds. In fact, assuming the asymptotic variance to be 1 for simplicity, as $n \to \infty$,

$$P(|\widehat{\theta}_n - \theta| > \varepsilon) \approx 2\mathcal{N}(-\varepsilon\sqrt{n}),$$

where $\mathcal{N}(\cdot)$ is the standard normal distribution function with

$$\mathcal{N}(-\varepsilon\sqrt{n}) = \frac{1}{\sqrt{2\pi}} \int_{-\infty}^{-\varepsilon\sqrt{n}} \mathrm{e}^{-\frac{x^2}{2}} dx \to 0 \quad \text{as } n \to \infty.$$

In contrast, the LDP results that we pursue in this book represent probabilistic errors as a function of n by deriving a rate function. With our large deviations results and the rate functions computed, for each n large enough, we have

$$P(|\widehat{\theta}_n - \theta| > \varepsilon) \leq K \exp(-c_0 n), \tag{1.1}$$

for some constant $K > 0$, where c_0 reflects the rate of exponential decay. When both upper and lower bounds are derived, they can be used to characterize guaranteed error bounds in probability using the upper bound and to study complexity issues using the lower bound, which offer distinctive aspects of identification errors beyond conventional measures of identification accuracy.

There exists a vast literature on the study of stochastic approximation and large deviations with emphasis on parameter estimations. Rates of convergence for stochastic approximation were studied in Kushner and Huang [30], which generalized an earlier result of Kushner for stochastic approximation and Monte Carlo methods. Their results were substantially extended in Kushner and Yin [31]. Large deviations for stochastic approximation algorithms were treated in Kushner [29] and Dupuis and Kushner [15]; large deviations lower and upper bounds were both obtained. In Dupuis and Kushner [16], escape probabilities for stochastic recursive approximation algorithms under general conditions were obtained. Concerning parameter estimations, the well-known work of Bahadur [2] and Bahadur et al. [3] (see also Kester and Kallenberg [27]) studied large deviations of estimators from a statistical point of view. Dealing with parameter estimation for the unknown parameter θ, these papers employed the inaccuracy rate

$$-\liminf_{n\to\infty} n^{-1} P_\theta(|T_n - \theta| > \varepsilon)$$

to measure the performance of a sequence of estimators $\{T_n\}$. They obtained lower bounds of the form

$$\liminf_{n\to\infty} n^{-1} P_\theta(|T_n - \theta| > \varepsilon) \geq -b(\varepsilon, \theta),$$

where $b(\cdot,\cdot)$ is defined from the Kullback–Leiber information $K(\eta, \theta)$ by $b(\varepsilon, \theta) = \inf\{K(\eta, \theta) : \eta \in \Theta, |\eta - \theta| > \varepsilon\}$. In [2], estimators for transition matrices of finite state Markov chains were developed. However, as mentioned in [3], the results cannot be extended to obtain upper bounds. Solo [49] studied H_∞ error bounds in the frequency domain for discrete-time systems $Y_k = G(q^{-1})u_k + U_k$, where q^{-1} is the back shift operator and G is a transfer function. He obtained lower bounds

$$\liminf_{n\to\infty} n^{-1} \log P(d(\widehat{g}_n(y), g(\theta_0)) > \varepsilon) \geq -b_{\theta_0},$$

where

$$\begin{aligned} b_{\theta_0}(\varepsilon) &= \inf_{\theta\in T}\{K(\theta, \theta_0) : d(g(\theta), g(\theta_0) > \varepsilon)\}, \\ K(\theta, \theta_0) &= \lim_{\theta} n^{-1} \log p_\theta(y^n)/p_{\theta_0}(y^n), \end{aligned}$$

are similar to these used in Bahadur's work. In [50], the large deviations results were presented in terms of a modulus of continuity that allows one to specify the probability of deviations.

In this book, the LDP on identification accuracy is developed for both regular sensors and quantized observations. The setup for identification under regular sensors follows the persistent identification framework introduced in [57], whereas system identification with binary and quantized sensors is developed within the quantized identification setup of [63]. Detailed treatments for uncorrelated noises are presented first. Due to their relative simplicity, explicit and tight error bounds are obtained and used in complexity analysis. On the other hand, to accommodate common practical scenarios of correlated noises, this book also includes mixing-type noises. Compared to the existing literature, we obtain the large deviations results in which the rate functions are explicitly calculated. In contrast to the traditional setup, unmodeled dynamics are included. We also deal with binary and quantized observations, which are needed in many applications. Using the rate functions obtained for the regular sensor, binary sensor, and quantized systems, we are able to compare the corresponding computational complexity. All three cases lead to exponential error bounds; the rate functions obtained enable us to pinpoint the different decaying rates of probabilistic errors. Further discussions on this and examples are presented in Section 4.4.

The rest of the book is arranged as follows. The system setup is given first in Chapter 2. Chapter 3 presents certain background materials on large deviations. Chapter 4 contains the main results on the LDP for systems identification under regular sensors, binary observations, and quantized data when the observation noises are independent and identically distributed random sequences. This simpler setup enables us to derive representations that reveal the main features of the identification problems. Also presented are some simulation examples to demonstrate the LDP on the estimates. The monotonicity of the rate functions under binary, quantized, and regular sensors is also discussed. Chapter 5 is devoted to large deviations of quantized identification under correlated noises.

Three case studies are presented in Chapters 6–8 to illustrate different aspects of LDPs in applications. Chapters 6 and 7 are applications of the LDP with regular sensors to diagnosis and complexity analysis, while Chapter 8 concerns quantized observations. Chapter 6 presents a battery diagnosis problem to demonstrate the utilization of the large deviations approach in deriving asymptotically accurate error characterization to guide diagnosis and prognosis reliability. In this study, in addition to the LDP rate function, the multiplication constant is also obtained from the underlying system models. This characterization can be used to determine the data size to ensure a targeted diagnosis accuracy. As another application, Chapter 7 focuses on signal processing for medical diagnosis of patients with lung or heart diseases. A cyclic blind signal separation algorithm is used to separate lung or heart sounds from noise, whose accuracy is directly linked to medical diagnosis reliability. To compare our signal/noise separation algorithms with the classical adaptive noise cancellation techniques and to evaluate algorithm performance under different noise features, we apply the LDPs to the benchmark examples for such studies. Chapter 8 presents a case of speed estimation for permanent magnet direct current (PMDC) motors with quantized sensors. Through the consideration of quantized sensing, the motor rotational speed can be estimated with low data flow rates. The LDP provides a probabilistic characterization of estimation accuracy to guide design of the sensing system and sampling rates to guarantee required estimation accuracy.

Chapter 9 describes how aperiodic inputs can be treated. It also explores the probabilities of the estimated trajectories escaping from a given domain in a continuous-time framework. Finally, it brings the book to a close with further remarks.

2

System Identification: Formulation

Consider a single-input–single-output (SISO) linear time-invariant (LTI) stable discrete-time system

$$y(t) = \sum_{i=0}^{\infty} a_i u(t-i) + d(t), \quad t = t_0 + 1, \ldots, \tag{2.1}$$

where $\{y(t)\}$ is the noise corrupted observation, $\{d(t)\}$ is the disturbance, $\{u(t)\}$ is the input with $u(t) = 0$ for $t < 0$, and $a = \{a_i, i = 0, 1, \ldots\}$, satisfying

$$\|a\|_1 = \sum_{i=0}^{\infty} |a_i| < \infty.$$

To proceed, we define

$$\begin{aligned} \theta &= (a_0, a_1, \ldots, a_{m_0-1})' \in \mathbb{R}^{m_0}, \\ \widetilde{\theta} &= (a_{m_0}, a_{m_0+1}, \ldots)', \end{aligned} \tag{2.2}$$

where z' denotes the transpose of z. Here θ is the vector-valued modeled part of the parameters, and $\widetilde{\theta}$ is known as the unmodeled dynamics. Separation of the modeled part and unmodeled dynamics is a standard modeling practice to limit model complexity [55, 57, 71], which enables us to treat parameters within a finite-dimensional space; see also related work [38] and references therein. This model complexity reduction produces model errors, due to the "truncation." Throughout this book, we assume that the input

Q. He et al., *System Identification Using Regular and Quantized Observations*, SpringerBriefs in Mathematics, DOI 10.1007/978-1-4614-6292-7_2,

u is uniformly bounded $\|u\|_\infty \le u_{\max}$, where $\|\cdot\|$ is the usual ℓ_1 norm. After applying u to the system and taking N output observations in the time interval $t_0, \ldots, t_0 + N - 1$, the observation can be rewritten as

$$y(t) = \varphi'(t)\theta + \widetilde{\varphi}'(t)\widetilde{\theta} + d(t), \quad t = t_0, \ldots, t_0 + N - 1, \tag{2.3}$$

where

$$\begin{aligned} \varphi(t) &= (u(t), u(t-1), \ldots, u(t - m_0 + 1))', \\ \widetilde{\varphi}(t) &= (u(t - m_0), u(t - m_0 - 1), \ldots)', \end{aligned}$$

or in vector form,

$$Y_N(t_0) = \Phi_N(t_0)\theta + \widetilde{\Phi}_N(t_0)\widetilde{\theta} + D_N(t_0),$$

where

$$\begin{aligned} Y_N(t_0) &= (y(t_0), \ldots, y(t_0 + N - 1))', \\ D_N(t_0) &= (d(t_0), \ldots, d(t_0 + N - 1))', \\ \Phi_N(t_0) &= (\varphi(t_0), \ldots, \varphi(t_0 + N - 1))', \\ \widetilde{\Phi}_N(t_0) &= (\widetilde{\varphi}(t_0), \ldots, \widetilde{\varphi}(t_0 + N - 1))'. \end{aligned} \tag{2.4}$$

Estimates will be derived from this relationship, depending on the sensing schemes used for y. Large deviations of the estimates will be investigated accordingly.

Remark 2.1 Typical linear finite-dimensional stable systems have rational transfer functions. When the systems are represented by impulse responses, they are always IIR (infinite impulse response), with exponentially decaying tails. Consequently, it is essential that the model structure (2.1) start with an IIR expression. Note that a truncation on the IIR is used to reach an FIR (finite impulse response) model plus an unmodeled dynamics. There are many other approaches to approximating a higher-order or infinite-dimensional system, including more general base functions, state space model reduction, Hankel-norm reduction, to name a few. It turns out that the FIR model allows much simpler algorithm development than other approaches. Consequently, we have adopted it here; see also [40] for discussion on discrete-time systems.

3
Large Deviations: An Introduction

The theory of large deviations characterizes probabilities and moments of certain sequences that are associated with "rare" events. In a typical application, consider the sum of N independent and identically distributed random variables. The deviations from the mean of the sum by a given bound become "rarer" as N becomes larger. Large deviations principles give asymptotically accurate probabilistic descriptions of such rare events as a function of N. Large deviations theory has been applied in diversified areas in probability theory, statistics, operations research, communication networks, information theory, statistical physics, financial mathematics, and queuing systems, among others.

The first rigorous large deviations results were obtained by Harald Cramér in the late 1930s, who used them to treat insurance risk models. He gave a large deviations principle (LDP) result for a sum of i.i.d. random variables, where the rate function is expressed by a power series. Then S.R.S. Varadhan developed a general framework for the LDP in 1966. Subsequently, large deviations problems were studied extensively by many researchers; see, for example, the work of Schilder for LDP of Brownian motions, Sanov for LDP of ergodic processes, and Freidlin and Wentzell for LDP of diffusions with some abstract foundation of LDP. A significant step forward was made through a series of papers by Donsker and Varadhan, starting in the mid 1970s, in which they developed a systematic large deviations theory for empirical measures of i.i.d. sequences and Markov processes. Their contributions were recognized, in particular by the award of the Abel Prize to Varadhan in 2007. In this chapter, we recall some of

Q. He et al., *System Identification Using Regular and Quantized Observations*, SpringerBriefs in Mathematics, DOI 10.1007/978-1-4614-6292-7_3,

the basic large deviations results. We are mainly concerned with a number of results that are particularly relevant to our work in this book, and the results presented in this chapter are far from comprehensive or exhaustive.

Consider one simple example. Let $X_1, \ldots, X_n$ be a sequence of i.i.d. random variables with finite mean $EX_1 = \mu$ and finite variance $E(X_1 - \mu)^2 = \sigma^2$. Then the law of large numbers states that the sample mean approaches the true mean as the sample size n goes to infinity. Define

$$S_n = \frac{1}{n}\sum_{k=1}^{n} X_k.$$

Then by the law of large numbers [17], as $n \to \infty$,

$$S_n \to \mu \quad \text{w.p.1}.$$

It is of great interest to obtain the related rate of convergence. To make this precise, let us fix a number $a > \mu$. By the law of large numbers, we know that $P(S_n > a) \to 0$ as n goes to infinity. In large deviations theory, one is interested in the rate at which the probability $P(S_n > a)$ decays to 0. To demonstrate the concepts, we define the following functions:

$$\begin{aligned} H(t) &= E[\mathrm{e}^{tX_1}], \\ I(\beta) &= \sup_t [t\beta - \log H(t)]. \end{aligned} \tag{3.1}$$

In the above, $H(t)$ is the moment-generating function. The function $I(\beta)$ is referred to as a Legendre–Fenchel transform and is also called a rate function. Note that $I(\beta)$ is always nonnegative. Then we formulate the first basic result of large deviations theory, which goes back to Cramér (1938). This result identifies the large deviations behavior of the empirical average S_n/n.

Theorem 3.1 *Let $\{X_n\}$ be a sequence of independent and identically distributed (i.i.d.) real-valued random variables satisfying*

$$H(t) = E\mathrm{e}^{tX_1} < \infty \quad \forall t \in \mathbb{R}.$$

Then for all $a > EX_1$,

$$\lim_{n\to\infty} \frac{1}{n} \log P(S_n \le an) = -I(a).$$

For the general definition of the large deviations principle, let $\{X^\varepsilon, \varepsilon > 0\}$ be a collection of random variables defined on a Polish space (i.e., a complete separable metric space) $(\Omega, \mathcal{F}, P)$ and taking values in a Polish space $\mathcal{E}$. Denote the metric on $\mathcal{E}$ by $d(\cdot,\cdot)$ and expectation with respect to P by E. The theory of large deviations focuses on random variables $\{X^\varepsilon\}$ for which the probabilities $P(X^\varepsilon \in A)$ converge to 0 exponentially fast for a class of Borel sets A. The exponential decaying rate of these probabilities is expressed in terms of a function I mapping $\mathcal{E}$ into $[0, \infty]$ (where $[0, \infty]$ denotes the extended nonnegative real numbers). This function is called a rate function if it has compact level sets, i.e., if for each $M < \infty$, the level set $\{x \in \mathcal{E} : I(x) \leq M\}$ is a compact subset in $\mathcal{E}$.

Definition 3.2 (Large deviations principle) Let I be a rate function on $\mathcal{E}$. The sequence $\{X^\varepsilon\}$ is said to satisfy the large deviations principle on $\mathcal{E}$ as $\varepsilon \to 0$ with rate function I if for every Borel set B in $\mathcal{E}$,

$$\begin{aligned} -\inf_{\beta \in B^\circ} I(\beta) &\leq \liminf_{\varepsilon \to 0} \varepsilon \log P\{X^\varepsilon \in B\} \\ &\leq \limsup_{\varepsilon \to 0} \varepsilon \log P\{X^\varepsilon \in B\} \\ &\leq -\inf_{\beta \in \overline{B}} I(\beta), \end{aligned} \tag{3.2}$$

where B° and $\overline{B}$ denote the interior and closure of B, respectively.

Remark 3.3 Instead of Definition 3.2, we can also define the following terms:

Large deviations upper bound: For each closed subset F of $\mathcal{E}$,

$$\limsup_{\varepsilon \to 0} \varepsilon \log P(X^\varepsilon \in F) \leq -\inf_{\beta \in F} I(\beta).$$

Large deviations lower bound: For each open subset G of $\mathcal{E}$,

$$\liminf_{\varepsilon \to 0} \varepsilon \log P(X^\varepsilon \in G) \geq -\inf_{\beta \in G} I(\beta).$$

Next we provide a couple of results on the LDP that will be used often in the following chapters. Proposition 3.4 is a version of the Gärtner–Ellis theorem (see [29, Lemma 1] and also [19]), which presents the LDP in the space $\mathbb{R}^k$. Here and hereinafter, we use $\langle a, b \rangle$ to denote the usual inner product in $\mathbb{R}^r$ for $a, b \in \mathbb{R}^r$ for a positive integer r.

Proposition 3.4 (Gärtner–Ellis theorem) *Let $\{X_k\}$ be a sequence of $\mathbb{R}^{m_0}$-valued random vectors for which the following limit exists:*

$$H(\tau) = \lim_{k \to \infty} \lambda_k \log E \exp\left\{\frac{1}{\lambda_k}\langle \tau, X_k \rangle\right\},$$

where $\tau \in \mathbb{R}^{m_0}$, $\lambda_k \to 0$, and $H(\cdot)$ is continuously differentiable. Define the dual function

$$I(\beta) = \sup_{\tau \in \mathbb{R}^{m_0}} [\langle \tau, \beta \rangle - H(\tau)].$$

Then $\{X_k\}$ satisfies the large deviations principle with rate function $I(\cdot)$.

The following statement confirms that the LDP is preserved under a continuous mapping, which is called the contraction principle; see [14, Theorem 4.2.1, p. 126]. In fact, the results in [14] are more general and can be applied to mappings between Hausdorff topological spaces. But the following assertion is sufficient for this book.

Proposition 3.5 (Contraction principle) *Let $f : \mathbb{R}^r \to \mathbb{R}^m$ be a continuous function. Assume that a family of random variables $\{X_k\}$ on $\mathbb{R}^r$ satisfies the LDP with speed $\{\lambda_k\}$ and rate function $I : \mathbb{R}^r \mapsto [0, +\infty]$. Then the family of random variables defined by $\{Y_k\}$, $Y_k = f(X_k)$, satisfies the LDP with speed $\{\lambda_k\}$ and rate function*

$$\widetilde{I}(y) = \inf \{I(x) : x \in \mathbb{R}^r,\ y = f(x)\}.$$

We conclude this chapter by providing an example. This illustrates the large deviations principle.

Example 3.6 Consider a sequence of i.i.d. random variables $\{X_n\}$ that follows the standard normal distribution. By the law of large numbers,

$$S_N = \frac{1}{N} \sum_{i=1}^{N} X_i \to 0 \quad \text{w.p.1.}$$

Given any number $0 < a$, the probability $P(S_N > a)$ tends to 0. Then it can be shown that

$$\lim_{N \to \infty} \frac{1}{N} \log P(S_N > a) = -I(a),$$

where $I(a) = a^2/2$. In fact, the rate function can be calculated directly by the Gärtner–Ellis theorem.

Example 3.7 (Schilder theorem) We consider a scaled random process

$$X^\varepsilon(t) = \varepsilon W(t), \ [0, T],$$

in $\mathbb{R}^n$, where $W(t)$ is a Wiener process in $\mathbb{R}^n$. It can be shown that as $\varepsilon \to 0$, the trajectory $X^\varepsilon(t)$ converges in probability to 0 on every finite time interval. How fast does the convergence take place? To answer this question, consider the Banach space $C_0 = C_0([0,T]; \mathbb{R}^n)$ of continuous functions equipped with the supremum norm $\|\cdot\|_\infty$. The process satisfies the large deviations principle with the rate function $I : C_0 \mapsto [0, \infty]$ given by

$$I(\varphi) = \begin{cases} \dfrac{1}{2} \displaystyle\int_0^T \dot{\varphi}(s) ds, & \text{if } \varphi \in C_0 \text{ is absolutely continuous,} \\ \infty, & \text{otherwise.} \end{cases}$$

In other words, for every open set $G \subseteq C_0$ and every closed set $F \subseteq C_0$,

$$\limsup_{\varepsilon \to 0} \varepsilon \log P(X^\varepsilon(\cdot) \in F) \leq - \inf_{\beta \in F} I(\beta)$$

and

$$\liminf_{\varepsilon \to 0} \varepsilon \log P(X^\varepsilon(\cdot) \in G) \geq - \inf_{\beta \in G} I(\beta).$$

4

LDP of System Identification under Independent and Identically Distributed Observation Noises

We first consider system identification under i.i.d. noise. Extension to correlated noises will be treated in Chapter 5. Beginning with the following assumptions, we should emphasize here that since we consider open-loop identification problems, the input signal u is part of experimental design and can be selected to enhance the identification process.

(A4.1) (a) $\{d(t)\}$ is a sequence of independent and identically distributed zero-mean random variables. Its moment-generating function $g(t)$ exists. (b) $\Phi_N(t_0)$ has full column rank. (c) The input signal $\{u(t)\}$ is periodic with period m_0. (d) $\|\widetilde{\theta}\|_1 \leq \widetilde{\eta}$.

Remark 4.1 Condition (b) is a persistent excitation condition, ensuring that the input is sufficiently rich for parameter estimation. When the input signal is periodic, this condition can be substantially simplified. Condition (c) is quite unique. In our previous work [57, 60, 63], we have demonstrated some key desirable features of periodic inputs: (1) under a periodic full-rank input, a complicated identification problem can be decomposed into a finite number of very simple identification problems; (2) under a periodic full-rank input, we can identify a rational transfer function under quantized observations without essential difficulties; (3) under an externally applied periodic full-rank input, we can identify a system in the closed-loop setting with guaranteed persistent excitation; (4) for applications of laws of large numbers, under periodic inputs, estimation errors can be written as direct averages, leading to the possibility of deriving not only upper bounds, but also CR (Cramér–Rao) lower bounds. However, aperiodic signals can

Q. He et al., *System Identification Using Regular and Quantized Observations*, SpringerBriefs in Mathematics, DOI 10.1007/978-1-4614-6292-7_4,

certainly be used. In this case, without the above benefits, we can still derive upper bounds on estimation errors under the persistent excitation condition (b). But the results will be more conservative, and lower bounds are harder to obtain. Since we aim to investigate complexity issues, tight error bounds are essential. Consequently, this book focuses on periodic full-rank inputs.

4.1 LDP of System Identification with Regular Sensors

Consider an m_0-periodic signal u, and set $N = km_0$ with an integer k. To simplify the expression, we write $\Phi_{m_0}(t_0)$ as Φ_0. We can write $\Phi_N(t_0)$ and $\widetilde{\Phi}_N(t_0)$ in Eq. (2.3) as

$$\begin{aligned}\Phi_N(t_0) &= (I_{m_0}, \ldots, I_{m_0})'\Phi_0,\\ \widetilde{\Phi}_N(t_0) &= (\Phi_N(t_0), \Phi_N(t_0), \ldots),\end{aligned}$$

where z' denotes the transpose of $z \in \mathbb{R}^{l_1 \times l_2}$ for $l_1, l_2 \geq 1$, and I_{m_0} denotes the $m_0 \times m_0$ identity matrix. In what follows, we apply the standard least squares estimation method. Define

$$L(t_0) = (\Phi_N'(t_0)\Phi_N(t_0))^{-1}\Phi_N'(t_0).$$

Then

$$\begin{aligned}L(t_0) &= (k\Phi_0'\Phi_0)^{-1}\Phi_0'(I_{m_0}, \ldots, I_{m_0})\\ &= \frac{1}{k}\Phi_0^{-1}(I_{m_0}, \ldots, I_{m_0}).\end{aligned} \tag{4.1}$$

Define the estimator $\widehat{\theta}_k = L(t_0)Y_N(t_0)$. Then

$$\begin{aligned}\widehat{\theta}_k &= L(t_0)(\Phi_N(t_0)\theta + \widetilde{\Phi}_N(t_0)\widetilde{\theta} + D_N(t_0))\\ &= \theta + L(t_0)\widetilde{\Phi}_N(t_0)\widetilde{\theta} + L(t_0)D_N(t_0).\end{aligned} \tag{4.2}$$

It follows that the deterministic part of the identification error becomes

$$\begin{aligned}\eta_k^d &= L(t_0)\widetilde{\Phi}_N(t_0)\widetilde{\theta}\\ &= \frac{1}{k}(\Phi_0^{-1}, \ldots, \Phi_0^{-1})(\Phi_N(t_0), \Phi_N(t_0), \ldots)\widetilde{\theta}\\ &= (I_{m_0}, I_{m_0}, \ldots)\widetilde{\theta}.\end{aligned} \tag{4.3}$$

Since η_k^d is independent of k, we write it as η^d. It is easily seen from Eq. (4.3) that $\|\eta^d\|_1 \leq \|\widetilde{\theta}\|_1$. The stochastic part of the identification error is

$$\begin{aligned} \eta_k^s &= L(t_0)D_N(t_0) \\ &= \frac{1}{k}\Phi_0^{-1}(I_{m_0}, \ldots, I_{m_0})D_N(t_0) \\ &= \Phi_0^{-1}(U_k^1, \ldots, U_k^{m_0})', \quad \text{where} \\ U_k^i &= \frac{1}{k}\sum_{l=0}^{k-1} d(t_0 + lm_0 + i), \quad \text{for} \quad i = 1, \ldots, m_0. \end{aligned} \tag{4.4}$$

Since $\{d(t)\}$ is an i.i.d. sequence, η_k^s tends to 0 w.p.1 as $k \to \infty$. Hence, $\lim_{k\to\infty} \widehat{\theta}_k = \theta + \eta^d$ w.p.1, where $\|\eta^d\|_1 \leq \|\widetilde{\theta}\|_1$. In Wang and Yin [57], identification error bounds were obtained using a combined approach of stochastic averaging and worst-case identification methods. To proceed, our task here is to establish probabilistic error bounds on η_k^s by the large deviations principle.

Remark 4.2 Typical probabilistic errors of identification problems consider only the form $P(|\widehat{\theta}_k - \theta| \geq \varepsilon)$ for some $\varepsilon > 0$. The LDP is a more general and refined property in that it permits probabilistic characterization of the estimates on any open or closed sets. For ease of presentation, the statement of the above theorem is concerned with an open set B. It can also be stated in terms of a closed set B. If a closed set B is used, we replace B by its interior B^0 and on the right-hand side replace $\overline{B}$ by B. In a more general setup, we can state the results using a Borel set B together with B^0 and $\overline{B}$ used on the left side and right side of the inequality, respectively.

To proceed, consider $U_k = (U_k^1, \ldots, U_k^{m_0})$ defined in Eq. (4.4) and let G be a linear function on $\mathbb{R}^{m_0}$ defined by $G(x) = \theta + \eta^d + \Phi_0^{-1}x$, $x \in \mathbb{R}^{m_0}$, where η^d comes from the unmodeled dynamics term. Note that $\widehat{\theta}_k = G(U_k)$ and Φ_0 is of full rank. To establish the LDP on $\widehat{\theta}_k$, we will first find the rate function of $\{U_k\}$ and then apply the contraction principle (Proposition 3.5) to derive the rate function of $\widehat{\theta}_k$. For $\tau = (\tau_1, \ldots, \tau_{m_0})' \in \mathbb{R}^{m_0}$ the H-functional of $\{U_k\}$ is

$$\begin{aligned} H(\tau) &= \lim_{k\to\infty} \frac{1}{k} \log E\exp\{k\langle U_k, \tau\rangle\} \\ &= \lim_{k\to\infty} \frac{1}{k} \log E\exp\{\sum_{l=0}^{k-1}\sum_{i=1}^{m_0} d(t_0 + lm_0 + i)\tau_i\} \\ &= \log E\exp\{\sum_{i=1}^{m_0} d(t_0 + i)\tau_i\} \\ &= \sum_{i=1}^{m_0} \log g(\tau_i). \end{aligned}$$

The Legendre transform of H is

$$I(\beta) = \sup_{\tau \in \mathbb{R}^{m_0}} [\langle \beta, \tau \rangle - H(\tau)] \\ = \sup_{\tau_1, \ldots, \tau_{m_0}} [\sum_{i=1}^{m_0} (\beta_i \tau_i - \log g(\tau_i))]. \tag{4.5}$$

Hence, by Proposition 3.4, $I(\beta)$ is the rate function of $\{U_k\}$.

Theorem 4.3 *Under assumption* (A4.1),

$$\widetilde{I}(\beta) = I(G^{-1}(\beta)) = I(\Phi_0(\beta - \theta - \eta^d))$$

is the rate function for $\{\widehat{\theta}_k\}$. *That is, for any open set* B *in* $\mathbb{R}^{m_0}$, *Eq.* (3.2) *holds with* $X_k = \widehat{\theta}_k$ *and* $\lambda_k = (1/k)$, *respectively.*

Proof. Since $\widehat{\theta}_k = G(U_k)$ and G is bijective and continuous, the result follows from the contraction principle (Proposition 3.5). □

Remark 4.4 If the i.i.d. noise $\{d(l)\}$ has the standard normal distribution, then the rate function $\widetilde{I}(\beta)$ has a simple form

$$\widetilde{I}(\beta) = \frac{|\Phi_0(\beta - \theta - \eta^d)|^2}{2}, \tag{4.6}$$

where $|\cdot|$ is the Euclidian norm.

4.2 LDP of System Identification with Binary Sensors

Suppose that the output y is measured by a binary sensor with a known threshold C. That is, we observe only

$$s(t) = \chi_{\{y(t) \le C\}} = \begin{cases} 1, & \text{if } y(t) \le C, \\ 0, & \text{otherwise.} \end{cases}$$

(A4.2) (a) $\{d(t)\}$ is a sequence of independent and identically distributed zero-mean random variables with distribution function $F(x)$ that is continuous and bijective and whose inverse F^{-1} exists and is continuous. The moment-generating function of $d(t)$ exists. (b) $\|\widetilde{\theta}\|_1 \le \widetilde{\eta}$.

In [63], Wang, Zhang, and Yin proposed an algorithm to determine θ, obtained its convergence, and studied the corresponding asymptotic distribution of normalized errors. Using the setup in Eq. (2.3), we recall the algorithm.

Algorithm:

Step 1. Define $Z_k = (Z_k^1, \ldots, Z_k^{m_0})'$ with

$$Z_k^i = \frac{1}{k} \sum_{l=0}^{k-1} s(t_0 + lm_0 + i), \quad i = 1, \ldots, m_0. \tag{4.7}$$

Note that the event $\{y(t_0 + lm_0 + i) \le C\}$ is the same as the event $\{d(t_0+lm_0+i) \le \widetilde{c}_i\}$, where $\widetilde{c}_i = C - \widetilde{C}_i$ and $\widetilde{C}_i$ is the ith component of $\Phi_0\theta + \widetilde{\Phi}_0\widetilde{\theta}$. Set $\widetilde{c} = (\widetilde{c}_1, \ldots, \widetilde{c}_{m_0})'$. Then, Z_k^i is the value of the k-sample empirical distribution.

Step 2. Since F is invertible, we can define

$$\begin{aligned} &\gamma_k^i = F^{-1}(Z_k^i),\ i = 1, \ldots, m_0 \ \text{ and} \\ &\gamma_k = (\gamma_k^1, \ldots, \gamma_k^{m_0})', \\ &L_k = C\mathbf{1} - \gamma_k,\ \text{ where } \mathbf{1} = (1, \ldots, 1)'. \end{aligned} \tag{4.8}$$

Step 3. When the input u is m_0-periodic and Φ_0 is invertible, we define the estimate by $\widehat{\theta}_k = \Phi_0^{-1} L_k$.

Remark 4.5 In [63], the following result was proved. Under assumption (A4.2), if the input u is m_0-periodic and Φ_0 is invertible, then $\widehat{\theta}_k \to \widehat{\theta}$ w.p.1 as $k \to \infty$. Furthermore, $\|\widehat{\theta} - \theta\|_1 \le \widetilde{\eta}$, where θ is the true vector-valued parameter, and $\widetilde{\eta} > 0$ is given in assumption (A4.2)(b), which represents the size of the unmodeled dynamics.

To proceed, define functions $\check{F}$ and $\check{F}^{-1}$ on $\mathbb{R}^{m_0}$ by

$$\begin{aligned} &\check{F}(v) = (F(v_1), \ldots, F(v_{m_0}))' \ \text{ and} \\ &\check{F}^{-1}(v) = (F^{-1}(v_1), \ldots, F^{-1}(v_{m_0}))' \text{ for } v \in \mathbb{R}^{m_0}. \end{aligned}$$

We now derive the large deviations principle for the identification problem. Define the function $\widehat{G} : \mathbb{R}^{m_0} \to \mathbb{R}^{m_0}$ by $\widehat{G}(x) = \Phi_0^{-1}[C\mathbf{1} - \check{F}^{-1}(x)]$ and write $\widehat{\theta}_k = \widehat{G}(Z_k)$. Since F and F^{-1} are bijective and continuous, we first study convergence rates of the sequence $\{Z_k\}$ and then apply the contraction principle, Proposition 3.5, to derive the rate function of the estimates $\{\widehat{\theta}_k\}$. First, we find the H-functional of $\{Z_k\}$,

$$\begin{aligned} H(\tau) &= \lim_{k\to\infty} \frac{1}{k} \log E \exp\{k\langle Z_k, \tau\rangle\} \\ &= \lim_{k\to\infty} \frac{1}{k} \log E \exp\{\sum_{l=0}^{k-1} \sum_{i=1}^{m_0} \chi_{\{d(t_0+lm_0+i)\le\widetilde{c}_i\}} \tau_i\} \\ &= \log E \exp\{\sum_{i=1}^{m_0} \chi_{\{d(t_0+i)\le\widetilde{c}_i\}} \tau_i\} \\ &= \sum_{i=1}^{m_0} \log[\mathrm{e}^{\tau_i} b_i + (1 - b_i)], \end{aligned}$$

where $b_i = P\{d(t_0) \leq \widetilde{c}_i\}$. The Legendre transform of H is given by

$$\begin{aligned} I(\beta) &= \sup_{\tau \in \mathbb{R}^{m_0}} [\langle \beta, \tau \rangle - H(\tau)] \\ &= \sup_{\tau_1,\ldots,\tau_{m_0}} [\sum_{i=1}^{m_0} (\beta_i \tau_i - \log(\mathrm{e}^{\tau_i} b_i + 1 - b_i))]. \end{aligned} \tag{4.9}$$

It is easily seen that $I(\beta) = \infty$ if there exists $1 \leq i \leq m_0$ such that $\beta_i < 0$ or $\beta_i > 1$. For $0 \leq \beta_i \leq 1$, $i = 1, 2, \ldots, m_0$, set

$$\overline{H}(\tau_1, \ldots, \tau_{m_0}) = \sum_{i=1}^{m_0} (\beta_i \tau_i - \log(\mathrm{e}^{\tau_i} b_i + 1 - b_i)).$$

To find $I(\beta)$, setting

$$\frac{\partial \overline{H}(\tau_1, \ldots, \tau_{m_0})}{\partial \tau_i} = \beta_i - \frac{\mathrm{e}^{\tau_i} b_i}{\mathrm{e}^{\tau_i} b_i + 1 - b_i} = 0, \quad \text{for} \quad i = 1, \ldots, m_0,$$

leads to

$$\tau_i^* = \log \frac{\beta_i (1 - b_i)}{b_i (1 - \beta_i)}, \quad 0 \leq \beta_i \leq 1. \tag{4.10}$$

Substituting Eq. (4.10) into Eq. (4.9) yields

$$\begin{aligned} I(\beta) &= \sup_{t \in \mathbb{R}^{m_0}} [\langle \beta, t \rangle - H(t)] \\ &= \langle \beta, t^* \rangle - H(t^*) \\ &= \sum_{i=1}^{m_0} \log \frac{\beta_i^{\beta_i} (1 - b_i)^{\beta_i - 1}}{b_i^{\beta_i} (1 - \beta_i)^{\beta_i - 1}}, \quad 0 \leq \beta_i \leq 1. \end{aligned}$$

To summarize,

$$I(\beta) = \begin{cases} \sum_{i=1}^{m_0} \log \frac{\beta_i^{\beta_i} (1 - b_i)^{\beta_i - 1}}{b_i^{\beta_i} (1 - \beta_i)^{\beta_i - 1}}, & 0 \leq \beta_i \leq 1, \ i = 1, \ldots, m_0, \\ \infty, \text{ otherwise.} & \end{cases}$$

Theorem 4.6 *The $I(\beta)$ given above is the rate function for the sequence $\{Z_k\}$ defined in Eq. (4.7). For any given open set B in $\mathbb{R}^{m_0}$, Eq. (3.2) holds with $\lambda_k = (1/k)$ and $X_k = Z_k$, respectively.*

Proof. A direct application of Proposition 3.4 leads to the result. □

We now derive explicit solutions to $\inf_{\beta \in \overline{B}} I(\beta)$. Without loss of generality, we assume $i = 1$. Consider the typical application in which we are interested in $P\{|Z_k - b_1| \geq \varepsilon\}$ for some small ε, namely, $\overline{B}_1 = (-\infty, b_1 - \varepsilon] \cup [b_1 + \varepsilon, \infty)$. Assume that $0 < b_1 - \varepsilon < b_1 + \varepsilon < 1$. Since the function

$$I(\beta_1) = \begin{cases} \log \frac{\beta_1^{\beta_1} (1 - b_1)^{\beta_1 - 1}}{b_1^{\beta_1} (1 - \beta_1)^{\beta_1 - 1}}, & 0 \leq \beta_1 \leq 1, \\ \infty, & \text{otherwise,} \end{cases}$$

reaches a minimum at b_1 and is monotone decreasing in $[0, b_1]$ and monotone increasing in $[b_1, 1]$, we have

$$\inf_{\beta_1 \in \overline{B_1}} I(\beta_1) = \begin{cases} \log \frac{(b_1+\varepsilon)^{(b_1+\varepsilon)}(1-b_1)^{b_1+\varepsilon-1}}{b_1^{(b_1+\varepsilon)}(1-(b_1+\varepsilon))^{(b_1+\varepsilon)-1}}, & b_1 \le 0.5, \\ \log \frac{(b_1-\varepsilon)^{(b_1-\varepsilon)}(1-b_1)^{b_1-\varepsilon-1}}{b_1^{(b_1-\varepsilon)}(1-(b_1-\varepsilon))^{(b_1-\varepsilon)-1}}, & b_1 > 0.5. \end{cases}$$

For small $\varepsilon > 0$ and both $b_1 < 0.5$ and $b_1 > 0.5$, $\inf_{\beta_1 \in \overline{B}_1} I(\beta_1)$ can be approximated using a Taylor expansion (with respect to ε) by the same expression

$$\inf_{\beta_1 \in \overline{B_1}} I(\beta_1) = \frac{\varepsilon^2}{2b_1(1-b_1)} + o(\varepsilon^2).$$

As a result, for small ε, the tail probability is dominated by

$$P\{|Z_k - b_1| \ge \varepsilon\} \le K\mathrm{e}^{-\frac{\varepsilon^2}{2b_1(1-b_1)}k}$$

for some $K > 0$. We point out that $E(s(1)-b_1)^2 = b_1(1-b_1)$ is the variance of the binary sequence and $b_1(1-b_1)/k$ is the Cramér–Rao lower bound in terms of mean squares estimation errors of the empirical measure (see [58]). For convenience, writing β_1 simply as β, Fig. 4.1 delineates variations on the rate function $I(\beta)$ under different values of p.

Theorem 4.7 *Under assumption* (A4.2),

$$\widehat{I}(\beta) = I(\widehat{G}^{-1}(\beta)) = I(\check{F}(C\mathbf{1} - \Phi_0\beta))$$

is the rate function for $\{\widehat{\theta}_k\}$. *That is, for any open set* B *in* $\mathbb{R}^{m_0}$, *Eq.* (3.2) *holds with* $X_k = \widehat{\theta}_k$ *and* $\lambda_k = (1/k)$, *respectively.*

Proof. Note that $\widehat{\theta}_k = \widehat{G}(Z_k)$, and $\widehat{G}$ and $\widehat{G}^{-1}$ are bijective and continuous. The result follows from Proposition 3.5. □

Remark 4.8 In fact, we can write the rate function $\widehat{I}(\cdot)$ explicitly as

$$\widehat{I}(\beta) = \sum_{i=1}^{m_0} \log \frac{F(C-(\Phi_0\beta)_i)^{F(C-(\Phi_0\beta)_i)}(1-b_i)^{F(C-(\Phi_0\beta)_i)-1}}{b_i^{F(C-(\Phi_0\beta)_i)}(1-F(C-(\Phi_0\beta)_i)^{F(C-(\Phi_0\beta)_i)-1}},$$

where $(\Phi_0\beta)_i$ denotes the ith component of $\Phi_0\beta$.

To illustrate, consider the case $m_0 = 1$. In this case, $F(C - \Phi_0\beta)$ is reduced to $F(C-u\beta)$, where $u \neq 0$ is a constant. Without loss of generality, assume $u = 1$. Now, to obtain $P\{|\widehat{\theta}_k - \theta| \ge \varepsilon\}$, we select for some small ε, $\overline{B}_1 = (-\infty, \theta - \varepsilon] \cup [\theta + \varepsilon, \infty)$. Let $\lambda = F(C - \beta)$. Since $F(\cdot)$ is a strictly monotonically increasing function, $\beta \le \theta - \varepsilon$ if and only if $\lambda \ge$

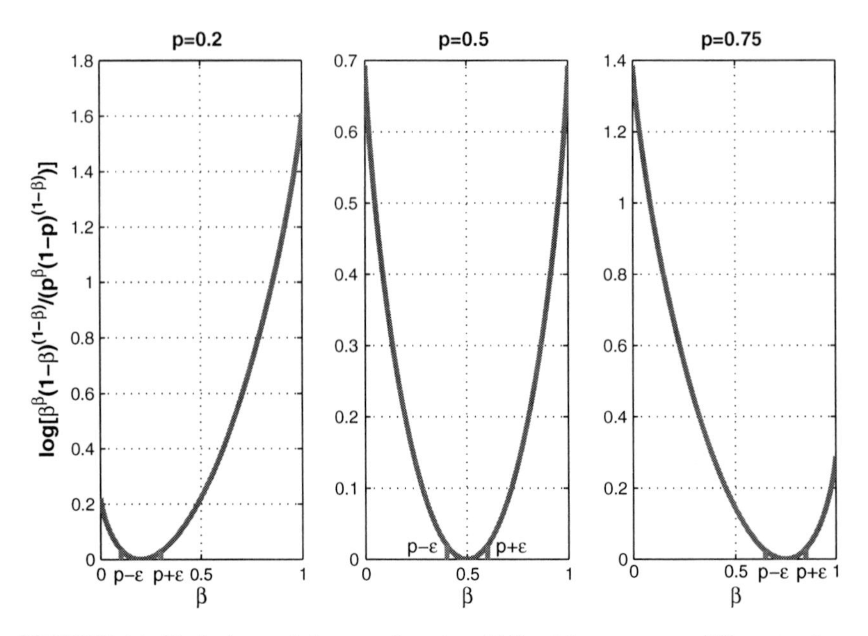

FIGURE 4.1. Variations of the rate function $I(\beta)$ with respect to different values of p

$F(C-\theta+\varepsilon)$, and $\beta \geq \theta+\varepsilon$ if and only if $\lambda \leq F(C-\theta-\varepsilon)$. Define $\widetilde{B}=(-\infty, F(C-\theta-\varepsilon)] \cup [F(C-\theta+\varepsilon), \infty)$. It follows that

$$\begin{aligned}
\inf_{\beta \in \overline{B}} \widehat{I}(\beta) &= \inf_{\lambda \in \widetilde{B}} I(\lambda) \\
&= \begin{cases} \log \frac{F(C-\theta+\varepsilon)^{F(C-\theta+\varepsilon)}(1-F(C-\theta))^{F(C-\theta+\varepsilon)-1}}{F(C-\theta)^{F(C-\theta+\varepsilon)}(1-F(C-\theta+\varepsilon))^{F(C-\theta+\varepsilon)-1}}, & C-\theta \leq 0, \\ \log \frac{F(C-\theta-\varepsilon)^{F(C-\theta-\varepsilon)}(1-F(C-\theta))^{F(C-\theta-\varepsilon)-1}}{F(C-\theta)^{F(C-\theta-\varepsilon)}(1-F(C-\theta-\varepsilon))^{F(C-\theta-\varepsilon)-1}}, & C-\theta > 0. \end{cases}
\end{aligned}$$

For small ε, $g(\varepsilon) := \inf_{\beta \in \overline{B}} \widehat{I}(\beta)$ has $g(0)=0$,

$$\dot{g}(0)=0, \quad \ddot{g}(0)=\frac{f^2(C-\theta)}{2F(C-\theta)(1-F(C-\theta))},$$

where $f(x)=\mathrm{d}F(x)/\mathrm{d}x$ is the probability density function. As a result, it can be approximated by

$$\inf_{\beta \in \overline{B}} \widehat{I}(\beta)=\frac{f^2(C-\theta)\varepsilon^2}{2F(C-\theta)(1-F(C-\theta))}+o(\varepsilon^2).$$

Hence asymptotically, the tail probability of the estimation error is dominated by

$$P\{|\theta_k-\theta| \geq \varepsilon\} \leq K \exp\left(-\frac{\varepsilon^2 f^2(C-\theta)}{2F(C-\theta)(1-F(C-\theta))}k\right)$$

for some $K > 0$. It is noted that

$$F(C-\theta)(1-F(C-\theta))/(kf^2(C-\theta))$$

is the Cramér–Rao lower bound (see [58]) in terms of mean squares estimation errors of $\widehat{\theta}_k$.

4.3 LDP of System Identification with Quantized Data

In this section, we study system identification under quantized observations, or equivalently sensors with multiple thresholds. For clarity and notational simplicity, we develop our results for the case $m_0 = 1$, namely a gain system. This assumption is not restrictive. General identification problems can be reduced to a set of identification problems for gains under periodic full-rank inputs; see [63] for details. Consider the gain system given by

$$y(l) = u(l)\theta + d(l), \quad l = 1, 2, \ldots,$$

where $u(l)$ is the input and $d(l)$ is the noise. The output $y(l)$ is measured by a sensor of m thresholds $-\infty < C_1 < \cdots < C_m < \infty$. The sensor can be represented by the indicator function $s(l) = (s^1(l), \ldots, s^m(l))'$, where $s^i(l) = \chi_{\{C_{i-1}<y(l)\le C_i\}}$, $i = 1, \ldots, m$, with $C_0 = -\infty$ and χ_A the indicator of the set A. Without loss of generality, assume $u(l) \equiv 1$ for all l. Then $y(l) = \theta + d(l)$. Under assumption (A4.2)(a), $\{y(l)\}$ is an i.i.d. sequence that has the accumulative distribution function $F(\cdot)$. Let

$$\begin{aligned} p_i &= P(C_{i-1} < y(l) \le C_i) \\ &= F(C_i - \theta) - F(C_{i-1} - \theta) \\ &:= F_i(\theta). \end{aligned}$$

Take k measurements on $s(l)$. Then

$$\xi_k^i = \frac{1}{k}\sum_{l=1}^{k} s^i(l)$$

for $i = 1, \ldots, m$ is the sample relative frequency of $y(l)$ taking values in $(C_{i-1}, C_i]$. It follows that ξ_k^i is an unbiased estimator of p_i for each k. An estimator θ_k^i of θ can be derived from $\xi_k^i = F_i(\theta_k^i)$. Define $G_i(x) = F_i^{-1}(x)$. Consequently, $\theta_k^i = G(\xi_k^i)$ is an estimator for θ. Define

$$\begin{aligned} &\Theta_k = (\theta_k^1, \ldots, \theta_k^m)', \\ &\xi_k = (\xi_k^1, \ldots, \xi_k^m)', \\ &G(v) = (G_1(v_1), \ldots, G_m(v_m))' \ \text{ for } \ v \in \mathbb{R}^m. \end{aligned}$$

It was shown in [58] that $\Theta_k = G(\xi_k)$ is an asymptotically unbiased estimator of $\theta\mathbf{1}$.

We are interested in the LDP on $\Theta_k \to \theta\mathbf{1}$. The H-functional of the sequence $\{\xi_k\}$ is

$$\begin{aligned} H(\tau) &= \lim_{k\to\infty} \frac{1}{k} \log E \exp\{k\langle \xi_k, \tau\rangle\} \\ &= \lim_{k\to\infty} \frac{1}{k} \log E \exp\{\sum_{i=1}^{m}\sum_{l=1}^{k} s^i(l) t_i\} \\ &= \log[\mathrm{e}^{\tau_1} p_1 + \mathrm{e}^{\tau_2} p_2 + \cdots + \mathrm{e}^{\tau_m} p_m + 1 - \sum_{i=1}^{m} p_i]. \end{aligned}$$

Define

$$\tau = (\tau_1, \ldots, \tau_m)', \; q(\tau) = (\mathrm{e}^{\tau_1}, \ldots, \mathrm{e}^{\tau_m})',$$
$$p = (p_1, \ldots, p_m)', \; \beta = (\beta_1, \ldots, \beta_m)'.$$

Assume $p_i > 0$ for $i = 1, \ldots, m$. Also, define $p_{m+1} = 1 - \mathbf{1}'p$ and $\beta_{m+1} = 1 - \mathbf{1}'\beta$. Then $H(\tau) = \log(p'q(\tau) + p_{m+1})$. The Legendre transform of H is given by

$$I(\beta) = \sup_{\tau\in\mathbb{R}^m} [\langle \beta, \tau\rangle - H(\tau)] = \sup_{\tau\in\mathbb{R}^m} [\beta'\tau - \log(p'q(\tau) + p_{m+1})].$$

Let $D = \{\beta : 0 \le \beta_i \le 1, \sum_{i=1}^m \beta_i \le 1\}$. If $\beta \notin D$, then $I(\beta) = \infty$. To see this, note that $\beta \notin D$ implies that β has either negative component or component greater than 1. Without loss of generality, let $\beta_1 < 0$. Then $I(\beta) = \infty$ by letting $\tau_i = -\infty$ for $i = 1, \ldots, m$. If $\beta_1 > 1$, we also get $I(\beta) = \infty$ by letting $\tau_1 = \infty$.

Next we consider the case of $\beta \in D$. To solve for the optimal τ^*, we consider the equation $\beta - \frac{\mathcal{P}q(\tau)}{p'q(\tau)+p_{m+1}} = 0$, where $\mathcal{P} = \mathrm{diag}(p)$. Note that $\mathbf{1}'\mathcal{P} = p'$ and $p'\mathcal{P}^{-1} = \mathbf{1}'$. It follows that $q(\tau^*) = (\mathcal{P} - \beta p')^{-1} p_{m+1}\beta$. By the matrix inversion lemma,

$$\begin{aligned} (\mathcal{P} - \beta p')^{-1} &= \mathcal{P}^{-1} + \mathcal{P}^{-1}\beta(1 - p'\mathcal{P}^{-1}\beta)^{-1} p'\mathcal{P}^{-1} \\ &= \mathcal{P}^{-1}(I + \frac{1}{\beta_{m+1}}\beta\mathbf{1}') \\ &= \frac{1}{\beta_{m+1}}\mathcal{P}^{-1}(\beta_{m+1} I + \beta\mathbf{1}'). \end{aligned}$$

This implies

$$\begin{aligned} q(\tau^*) &= \frac{1}{\beta_{m+1}}\mathcal{P}^{-1}(\beta_{m+1} I + \beta\mathbf{1}') p_{m+1}\beta \\ &= \frac{p_{m+1}}{\beta_{m+1}}\mathcal{P}^{-1}\beta, \end{aligned}$$

and

$$\tau^* = \log\frac{p_{m+1}}{\beta_{m+1}}\mathbf{1} + \log(\beta_1/p_1, \ldots, \beta_m/p_m)'.$$

In addition,

$$\begin{aligned} p'q(\tau^*) + p_{m+1} &= \frac{p_{m+1}}{\beta_{m+1}} p'\mathcal{P}^{-1}\beta + p_{m+1} \\ &= \frac{p_{m+1}}{\beta_{m+1}} \mathbf{1}'\beta + p_{m+1} \\ &= \frac{p_{m+1}}{\beta_{m+1}}. \end{aligned}$$

Consequently, $I(\beta)$ has the explicit expression

$$\begin{aligned} I(\beta) &= \beta' \log \frac{p_{m+1}}{\beta_{m+1}} (\beta_1/p_1, \ldots, \beta_m/p_m)' - \log \frac{p_{m+1}}{\beta_{m+1}} \\ &= \mathbf{1}'\beta \log \frac{p_{m+1}}{\beta_{m+1}} - \log \frac{p_{m+1}}{\beta_{m+1}} + \sum_{i=1}^{m} \beta_i \log \frac{\beta_i}{p_i} \\ &= \sum_{i=1}^{m+1} \beta_i \log \frac{\beta_i}{p_i}. \end{aligned}$$

To summarize,

$$I(\beta) = \begin{cases} \sum_{i=1}^{m+1} \beta_i \log \frac{\beta_i}{p_i}, & \beta \in D, \\ \infty, \text{ otherwise.} \end{cases} \tag{4.11}$$

Remark 4.9 Note that the binary-sensor case and quantized-sensor case can also be derived from Sanov's theorem of empirical measures (see [14, Theorem 2.1.10, p. 16]). The Sanov theorem states that the rate function of empirical measures is the relative entropy. In our case, it coincides with our calculation. For any probability vectors $v, \widetilde{v} \in \mathbb{R}^{m+1}$, i.e., $v_i, \widetilde{v}_i \geq 0$ and

$$\sum_{i=1}^{m+1} v_i = 1, \ \sum_{i=1}^{m+1} \widetilde{v}_i = 1,$$

the relative entropy is defined by

$$H(v|\widetilde{v}) = \sum_{i=1}^{m+1} v_i \log(\frac{v_i}{\widetilde{v}_i}).$$

By the entropy form, the rate function is $I(\beta) = H(\beta|p)$ if we define

$$\beta = (\beta_1, \ldots, \beta_{m+1})' \ \text{ and } \ p = (p_1, \ldots, p_{m+1})'.$$

In fact, for the binary case, we can write

$$\begin{aligned} I(\beta) &= \sum_{i=1}^{m_0} \log \frac{\beta_i^{\beta_i}(1-b_i)^{\beta_i - 1}}{b_i^{\beta_i}(1-\beta_i)^{\beta_i - 1}} \\ &= \sum_{i=1}^{m_0} \left(\beta_i \log \frac{\beta_i}{b_i} + (1-\beta_i) \log \frac{1-\beta_i}{1-b_i} \right) \\ &= \sum_{i=1}^{m_0} H(\widetilde{\beta}_i | \widetilde{b}_i), \end{aligned}$$

if we define $\widetilde{\beta}_i = (\beta_i, 1-\beta_i)', \widetilde{b}_i = (b_i, 1-b_i)'$.

We now establish a monotonicity property in terms of the number of thresholds in quantized observations. This may be viewed as a partition property of relative entropies. Suppose that starting from the existing thresholds $\{C_1, \ldots, C_m\}$, one additional threshold $\widetilde{C}$ is added. Without loss of generality, assume $C_{i-1} < \widetilde{C} < C_i$. As a result, $p_i > 0$ is decomposed into two probabilities:

$$\begin{aligned} p_i^1 &= P(C_{i-1} < x \le \widetilde{C}), \\ p_i^2 &= P(\widetilde{C} < x \le C_i), \end{aligned}$$

with $p_i^1 + p_i^2 = p_i$. Excluding the trivial cases, we assume $0 < p_i^1 < p_i$. Similarly, β_i is decomposed into

$$\beta_i = \beta_i^1 + \beta_i^2, \quad 0 < \beta_i^1 < \beta_i.$$

This refinement on the threshold set expands

$$\beta = (\beta_1, \ldots, \beta_i, \ldots, \beta_{m+1})'$$

and

$$p = (p_1, \ldots, p_i, \ldots, p_{m+1})'$$

to

$$\begin{aligned} \widetilde{\beta} &= (\beta_1, \ldots, \beta_i^1, \beta_i^2, \ldots, \beta_{m+1})' \text{ and} \\ \widetilde{p} &= (p_1, \ldots, p_i^1, p_i^2, \ldots, p_{m+1})'. \end{aligned}$$

Lemma 4.10 $I(\widetilde{\beta}) \ge I(\beta)$.

Proof. Consider

$$f(\beta_i^1) := \beta_i^1 \log \frac{\beta_i^1}{p_i^1} + (\beta_i - \beta_i^1) \log \frac{\beta_i - \beta_i^1}{p_i - p_i^1}.$$

From Eq. (4.11), we need only show that for all β_i^1 satisfying $0 \le \beta_i^1 \le \beta_i$,

$$f(\beta_i^1) \ge \beta_i \log \frac{\beta_i}{p_i}.$$

First, on the boundaries of $[0, \beta_i]$,

$$\begin{aligned} f(0) &= \beta_i \log \frac{\beta_i}{p_i - p_i^1} \geq \beta_i \log \frac{\beta_i}{p_i}; \\ f(\beta_i) &= \beta_i \log \frac{\beta_i}{p_i^1} \geq \beta_i \log \frac{\beta_i}{p_i}. \end{aligned}$$

In the interior, the condition

$$\frac{\mathrm{d} f(\beta_i^1)}{\mathrm{d}\beta_i^1} = \log \frac{\beta_i^1}{p_i^1} - \log \frac{\beta_i - \beta_i^1}{p_i - p_i^1} = 0$$

leads to the stationary point

$$\frac{\widehat{\beta}_i^1}{p_i^1} = \frac{\beta_i}{p_i},$$

and

$$f(\widehat{\beta}_i^1) = \beta_i \log \frac{\beta_i}{p_i}.$$

Since

$$\frac{\mathrm{d}^2}{\mathrm{d}\beta^2} f(\widehat{\beta}_i^1) = \frac{1}{\widehat{\beta}_i^1} + \frac{1}{\beta_i - \widehat{\beta}_i^1} > 0,$$

$\widehat{\beta}_i^1$ is indeed a minimum point. As a result, $\inf_{0<\beta_i^1<\beta_i} f(\beta_i^1) = \beta_i \log \frac{\beta_i}{p_i}$, which implies the desired inequality. □

Applying the Gärtner–Ellis theorem (Proposition 3.4), we obtain that $I(\cdot)$ is the rate function for the sequence $\{\xi_k\}$. Define $C = (C_1, \ldots, C_m)'$ and $\widehat{F}(v) = (F_1(v_1), \ldots, F_m(v_m))'$ for $v \in \mathbb{R}^m$. By virtue of Proposition 3.5, the rate function of Θ_k is

$$\widetilde{I}(\widehat{\beta}) = \inf\{I(\beta) : G(\beta) = \widehat{\beta}\} = I(\widehat{F}(\widehat{\beta})). \tag{4.12}$$

Since Θ_k converges to the vector $\theta\mathbf{1}$, we may construct an estimator $\widehat{\theta}_k$ of θ by $\widehat{\theta}_k = \sum_{i=1}^m \gamma_i \theta_k^i = \gamma' \Theta_k$, where $\gamma = (\gamma_1, \ldots, \gamma_m)$ with $\gamma_1 + \cdots + \gamma_m = 1$ and $\widehat{\theta}_k$ is asymptotically unbiased.

To find the LDP for the estimator $\widehat{\theta}_k$, we apply Proposition 3.5 to derive the rate function $I_\gamma(\beta) = \inf\{\widetilde{I}(\widehat{\beta}) : \gamma' \widehat{F}(\widehat{\beta}) = \beta\}$. For the typical case of characterizing the error probability $P\{|\widehat{\theta}_k - \theta| \geq \varepsilon\}$, the set of interest is $B = (-\infty, \theta - \varepsilon] \cup [\theta + \varepsilon, \infty)$. To calculate $\widehat{I}(\varepsilon) = \inf_{\beta \in \overline{B}} I_\gamma(\beta)$, we solve two constrained optimization problems:

$$\begin{aligned} \widehat{I}^+(\varepsilon) &= \inf \sum_{i=1}^{m+1} \beta_i \log \frac{\beta_i}{p_i} \\ \text{s.t.} \quad & 0 < \beta_i < 1, \quad i = 1, \ldots, m, \\ & \sum_{i=1}^{m+1} \beta_i = 1 \;\; \text{and} \;\; \sum_{i=1}^{m} \gamma_i F_i(\beta_i) \geq \theta + \varepsilon, \end{aligned}$$

$$\begin{aligned}
\widehat{I}^-(\varepsilon) &= \inf \sum_{i=1}^{m+1} \beta_i \log \frac{\beta_i}{p_i} \\
\text{s.t.} \quad & 0 < \beta_i < 1, \quad i = 1, \ldots, m, \\
& \sum_{i=1}^{m+1} \beta_i = 1 \text{ and } \sum_{i=1}^{m} \gamma_i F_i(\beta_i) \leq \theta - \varepsilon.
\end{aligned}$$

Then $\widehat{I}(\varepsilon) = \min\{\widehat{I}^+(\varepsilon), \widehat{I}^-(\varepsilon)\}$. Although we do not expect to obtain closed-form solutions generally, the above constrained optimization problem characterizes the desired solution.

4.4 Examples and Discussion

We will present several simulation examples to demonstrate the large deviations principle for the estimates and illustrate complexity relationships among binary, quantized, and regular sensors by proving the monotonicity property of the corresponding rate functions.

Binary sensors. We shall start with system identification under binary sensors. In this example, $\theta' = (1.75, 1.75, 2.75)$ is the true parameter vector, and for simplicity, assume $\widetilde{\theta} = 0$, i.e., no unmodeled dynamics. Select a 3-periodic input with valued for one period of $(u(1), u(2), u(3)) = (3, 4, 5)$, which is of full rank, and

$$\Phi_0 = \begin{bmatrix} 3 & 4 & 5 \\ 4 & 5 & 3 \\ 5 & 3 & 4 \end{bmatrix}.$$

The noise is an i.i.d. sequence of random variables with the standard normal distribution. The binary sensor has a threshold $C = 25$. For $\varepsilon = 1$, we compare empirical measures of $P(|\widehat{\theta}_k - \theta| > \varepsilon)$, with the calculated large deviations bound

$$\exp\left(-k \inf_{|x-\theta|>\varepsilon} I(x)\right).$$

Step 1. For each $k = 5, \ldots, 50$, the identification algorithm is repeated 1000 times, and the sample frequencies of the event

$$|\widehat{\theta}_k - \theta| > \varepsilon$$

are then calculated as an approximation to

$$P(|\widehat{\theta}_k - \theta| > \varepsilon).$$

Step 2. From Remark 4.8, compute

$$\widehat{I}(\beta) = \sum_{i=1}^{m_0} \log \frac{F(C - (\Phi_0\beta)_i)^{F(C-(\Phi_0\beta)_i)} (b_i - 1)^{F(C-(\Phi_0\beta)_i)-1}}{b_i^{F(C-(\Phi_0\beta)_i)} (F(C - (\Phi_0\beta)_i) - 1)^{1-F(C-(\Phi_0\beta)_i)}},$$

where

$$b_i = P\{d(l) \le C - (\Phi_0\theta)_i\}.$$

It is calculated that $\inf_{|x-\theta|>\varepsilon} I(x) = -0.16$, and the LDP exponentially decaying curve is $\exp(-0.16k)$. These are shown in Fig. 4.2. For sufficiently large k, these two curves match very well.

Regular sensors. We consider the same example of system identification under a regular sensor with $\varepsilon = 0.5$. In this case, we aim to find the convergence rate of

$$\begin{aligned}\widehat{\theta}_k &= \theta + \eta_s^k \\ &= \theta + \frac{1}{k}\Phi_0^{-1}(I_3, I_3, I_3)D_{3k} \\ &= \theta + \Phi_0^{-1}(U_k^1, U_k^2, U_k^3)',\end{aligned}$$

where

$$U_k^j = \frac{1}{k}\sum_{l=0}^{k-1} d(t_0 + lm_0 + j), \quad \text{for} \quad j = 1, 2, 3,$$

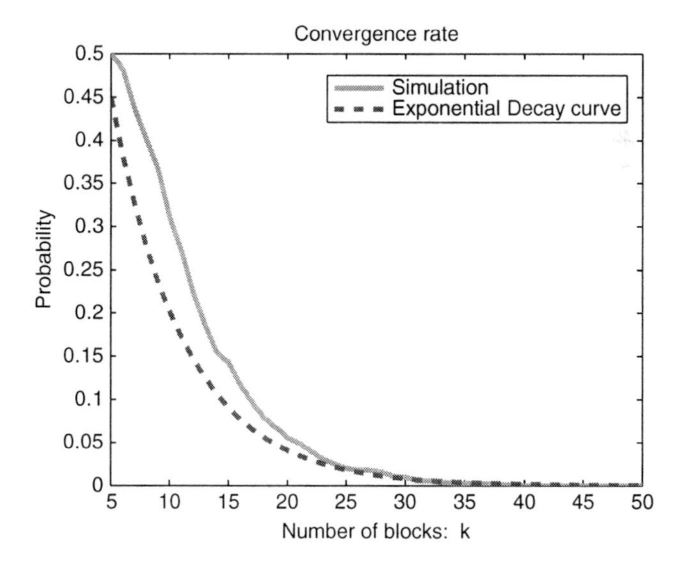

FIGURE 4.2. Comparison of the empirical errors and the LDP bound under a binary sensor

and Φ_0 is as in the last example. By virtue of Theorem 4.3, the rate function is

$$\widetilde{I}(\beta) = I(G^{-1}(\beta)) = I(\Phi_0\beta - \theta),$$

where

$$I(\beta) = \sup_{\tau_1,\ldots,\tau_{m_0}} \Big[\sum_{i=1}^{m_0} (\beta_i\tau_i - \log g(\tau_i))\Big],$$

and $g(\cdot)$ is the moment-generating function given in (A4.1). Since the noise d_l is a standard normal random variable, we obtain

$$I(\beta) = \frac{|\beta - \theta|^2}{2}$$

by Remark 4.4. Hence,

$$\widetilde{I}(\beta) = I(G^{-1}(\beta)) = \frac{|\Phi_0\beta - \theta|^2}{2}.$$

We simulate the probability

$$P(|\widehat{\theta}_k - \theta| > 0.5),$$

and then compare with the large deviations result, in which this probability is approximated by

$$K \exp\big(-k \inf_{|x|>1} \widetilde{I}(x)\big)$$

for some $K > 0$.

Step 1. For each $k = 5, \ldots, 100$, take 1000 samples and use the proportion of the number of times of

$$|\widehat{\theta}_k - \theta| > 0.5$$

to approximate

$$P(|\widehat{\theta}_k - \theta| > 0.5).$$

Step 2. By calculating the value

$$\inf_{|x-\theta|>0.5} \widetilde{I}(x) = \inf_{|x-\theta|>0.5} \frac{|\Phi_0(x-\theta)|^2}{2} = 0.375,$$

the exponentially decaying curve is $\exp(-0.375k)$ (Fig. 4.3).

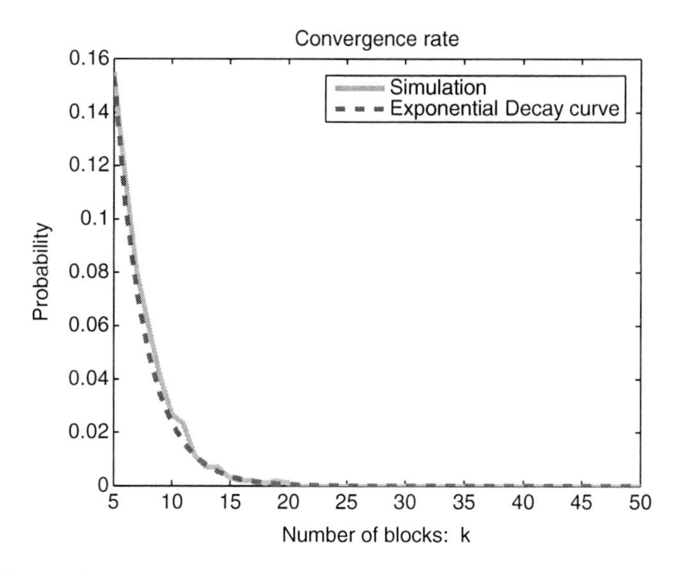

FIGURE 4.3. Comparison of the empirical errors and the LDP bound under a regular sensor

4.4.1 Space Complexity: Monotonicity of Rate Functions with Respect to Numbers of Sensor Thresholds

The LDP indicates that

$$P(|\widehat{\theta}_k - \theta| \geq \varepsilon) \leq K \exp(- \inf_{|\beta-\theta|\geq\varepsilon} I^0(\beta)k),$$

for some $K > 0$, where I^0 is the rate function depending on the sensor types. Comparison of the rate functions under different sensor types will demonstrate complexity and benefits relationships in sensor design. Let $y(l) = \theta + d(l)$ for $l = 1, 2, \ldots$, where $d(l)$ is a sequence of i.i.d. noise with the standard normal distribution. We assume that $\theta = 3$, $C_1 = 2$, and $C_2 = 4$. We want to find the probability

$$P(|\widehat{\theta}_k - \theta| > 1).$$

a. Observations under binary sensors. Under a binary sensor of threshold $C_1 = 2$, we have the estimator

$$\widehat{\theta}^b_k = C_1 - F^{-1}(\xi^1_k),$$

where

$$\xi^1_k = \frac{\sum_{l=1}^k \chi_{\{d(l)\leq C_1-\theta\}}}{k}.$$

By Remark 4.8, $\widehat{\theta}_k^b$ has the rate function

$$I^b(\beta) = \log \frac{F(C_1-\beta)^{F(C_1-\beta)}(b_1-1)^{F(C_1-\beta)-1}}{b_1^{F(C_1-\beta)}(F(C_1-\beta)-1)^{1-F(C_1-\beta)}},$$

where $b_1 = P(d(l) \leq C_1 - \theta)$. Hence the estimator

$$\Theta_k^b = (\widehat{\theta}_k^b, \widehat{\theta}_k^b)'$$

has the rate function

$$I^b(\widehat{\beta}) = \inf\{I^b(\beta) : (\beta,\beta)' = \widehat{\beta}\}.$$

Applying the LDP and the contraction principle Proposition 3.5, with $\Theta = (\theta,\theta)'$, we obtain

$$\begin{aligned}\lim_{k\to\infty} \frac{1}{k} \log P(|\Theta_k^b - \Theta| > 1) &= - \inf_{|\widehat{\beta}-\Theta|>1} I^b(\widehat{\beta}) \\ &= - \inf_{(\beta-3)^2 > \frac{1}{2}} I^b(\beta) = -0.0658.\end{aligned}$$

b. Observations under quantized sensors. Consider a quantized sensor with two thresholds $C_1 = 2$ and $C_2 = 4$. The estimator for $\Theta = (3,3)'$ is

$$\Theta_k^q = (C_1 - F^{-1}(\xi_k^1), C_2 - F^{-1}(\xi_k^2))',$$

where

$$\xi_k^i = \sum_{l=1}^{k} \chi_{\{d(l)\leq C_i-\theta\}}/k$$

for $i = 1, 2$. By Eq. (4.12), the rate function for Θ_k^q is

$$\begin{aligned}I^q((\beta_1,\beta_2)') &= \sup_{\tau_1,\tau_2} [\tau_1 F(C_1-\beta_1) \\ &\quad + \tau_2 F(C_2-\beta_2) - H(\tau_1,\tau_2)],\end{aligned}$$

where

$$H(\tau_1,\tau_2) = \log[\mathrm{e}^{\tau_1+\tau_2} p_1 + \mathrm{e}^{\tau_2}(p_2-p_1) + (1-p_2)],$$

and $p_i = P\{d(l) \leq C_i - \theta\}$, $i = 1, 2$. Applying the large deviations principle, we obtain

$$\lim_{k\to\infty} \frac{1}{k} \log P(|\Theta_k^q - \Theta| > 1) = - \inf_{|\widehat{\beta}-\Theta|>1} I^q(\widehat{\beta}) = -0.1014.$$

c. Observations under regular sensors. When we use regular sensors, the estimator of θ is given by

$$\widehat{\theta}_k^r = \frac{\sum_{l=1}^k y_l}{k}.$$

By Remark 4.4, the rate function of $\widehat{\theta}_k^r$ is $I^r(\beta) = (\beta - 3)^2/2$. Hence the rate function of the two-dimensional estimator

$$\Theta_k^r = (\widehat{\theta}_k^r, \widehat{\theta}_k^r)'$$

is

$$I^r(\widehat{\beta}) = \inf\{I^r(\beta) : (\beta, \beta)' = \widehat{\beta}\}.$$

Applying the LDP and contraction principle Proposition 3.5, we have

$$\begin{aligned}\lim_{k\to\infty} \frac{1}{k} \log P(|\Theta_k^r - \Theta| > 1) &= - \inf_{|\widehat{\beta}-\Theta|>1} I^r(\widehat{\beta}) \\ &= - \inf_{(\beta-3)^2>\frac{1}{2}} \frac{(\beta-3)^2}{2} = -0.25.\end{aligned}$$

From the above discussion of the three different sensors, it is clear that there is a monotonicity of the rate functions when the sensor complexity increases, which can be summarized as

$$P(|\theta_k - \theta| > 1) = \begin{cases} \exp(-0.0658k) \text{ binary sensor,} \\ \exp(-0.1014k) \text{ quantized sensor,} \\ \exp(-0.25k) \text{ regular sensor,} \end{cases} \tag{4.13}$$

where the quantized sensor has two threshold values. Figure 4.4 displays the comparison results. In view of the study in [61], the two-threshold quantized sensor design is a refinement of the binary sensor case, and the regular sensor is an "infinite" refinement of quantized sensors. The large deviations rate functions give precise descriptions of convergence rates, hence can be used in selecting sensor complexity levels.

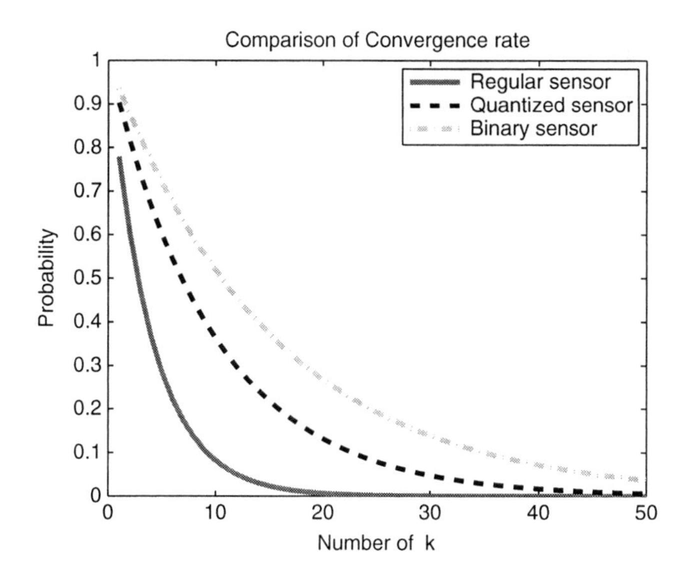

FIGURE 4.4. Comparison of convergence rates among different sensors

5

LDP of System Identification under Mixing Observation Noises

Up to this point, the observation noises are assumed to be uncorrelated. In this chapter, we demonstrate that a much larger class of noise processes can be treated.

5.1 LDP for Empirical Means under ϕ-Mixing Conditions

It is natural to consider the noise $\{d(l)\}$ under mixing conditions. In this section we consider stationary ϕ-mixing random sequences with sufficiently fast convergence rates. Let $\{X_k\}$ be a stationary sequence. Denote by $\mathcal{F}_a^b$ the σ-algebra generated by $\{X_k : a \leq k \leq b\}$, and let

$$\phi(k) = \sup\{|P(B|A) - P(B)| : A \in \mathcal{F}_0^l, P(A) > 0, B \in \mathcal{F}_{k+l}^{\infty}, l \in Z_+\}.$$

$\{X_k\}$ is said to be ϕ-mixing if $\phi(k) \to 0$ as $k \to \infty$. Throughout this section we will need the following assumption.

(A5.1) (a) $\{X_k\}$ is a stationary ϕ-mixing random sequence and $\phi(k)$ satisfies

$$\phi(k) \leq \exp(-kr(k)),$$

where $r(k) \to \infty$ as $k \to \infty$, and

$$\sum_{k=1}^{\infty} \frac{r(k)}{k(k+1)} < \infty.$$

Q. He et al., *System Identification Using Regular and Quantized Observations*, SpringerBriefs in Mathematics, DOI 10.1007/978-1-4614-6292-7_5,

(b) $\{X_k\}$ takes values in a compact set $K_c \subset \mathbb{R}^d$.

Mixing processes are those whose remote past and distant future are asymptotically independent. For general reference on mixing processes, we refer the reader to [32]. Under assumption (A5.1), we derive the following theorem.

Theorem 5.1 *If* $\{X_k\}$ *is a stationary* ϕ*-mixing sequence satisfying* (A5.1), *then*

$$\widehat{S}_k = \frac{X_1 + X_2 + \cdots + X_k}{k}, \quad k \geq 1,$$

satisfies the LDP. That is, there is a rate function $I : \mathbb{R}^d \to [0, \infty]$ *that is convex and lower semicontinuous such that for every* $B \subset \mathbb{R}^d$,

$$\begin{aligned} -\inf_{\gamma \in B^\circ} I(\gamma) &\leq \liminf_{k \to \infty} \frac{1}{n} \log P\{\widehat{S}_k \in B\} \\ &\leq \limsup_{k \to \infty} \frac{1}{k} \log P\{\widehat{S}_k \in B\} \\ &\leq -\inf_{\gamma \in \overline{B}} I(\gamma), \end{aligned}$$

where B° *and* $\overline{B}$ *denote the interior and closure of* B, *respectively. Moreover, the rate function is given as*

$$I(\gamma) = \sup_{\beta \in \mathbb{R}^d} [\langle \gamma, \beta \rangle - \Lambda(\beta)], \tag{5.1}$$

where

$$H(\beta) = \lim_{k \to \infty} k^{-1} \log E \exp(k \langle \beta, \widehat{S}_k \rangle).$$

Remark 5.2 Dealing with empirical measures, a large deviations principle for a class of stationary processes under certain mixing conditions was proved in [14]. Their condition is different from ours and is implied by the so-called ψ-mixing processes. The large deviations principle for arithmetic means of a ϕ-mixing process under assumption (A5.1) without $\sum_{k=1}^{\infty} \frac{r(k)}{k(k+1)} < \infty$ was proved in [7]. However, the proof is a bit complicated. Here, we use an approach similar to that of [14] to derive an alternative proof of the LDP for a sample mean under ϕ-mixing assumptions. Our results on the large deviations for the identification are then based on the large deviations for the sample mean. In order to prove the desired result, we need a number of preparatory results. They are stated in the following proposition. The first part in the proposition is the approximate subadditivity [14, Lemma 6.4.10, p.282], and the second and third parts are in [14, Lemma 4.4.8, Theorem 4.4.10, p.143], respectively.

Proposition 5.3 *The following results hold.*

(a) *Let $h : \mathbb{N} \to \mathbb{R}$ and assume that all $k, l \geq 1$,*

$$h(k+l) \leq h(k) + h(l) + \varepsilon(k+l),$$

where $\varepsilon(k)$ is a nondecreasing sequence satisfying

$$\sum_{k=1}^{\infty} \frac{\varepsilon(k)}{k(k+1)} < \infty.$$

Then

$$\lim_{k\to\infty} \frac{h(k)}{k}$$

exists and is finite.

(b) *Let $\{X_k\}$ be a sequence of random variables taking values in a compact subset $K_c \subset \mathbb{R}^d$, and for every concave, bounded above, and continuous function g, we have $\Lambda_g = \lim_{k\to\infty} \log E e^{kg(X_k)}$. Then Λ_f exists for all f belonging to $C_b(\mathbb{R}^d)$, which is the space of all bounded and continuous functions on $\mathbb{R}^d$. Furthermore, $\{X_k\}$ satisfies the LDP with the rate function*

$$I(\beta) = \sup_{f \in C_b(\mathbb{R}^d)} [f(\beta) - H_f].$$

(c) *Assume that $\{X_k\}$ satisfies the LDP with a rate function $I(\cdot)$ that is convex, and $\limsup_{k\to\infty} \frac{1}{k} \log E e^{k\langle \tau, X_k \rangle} < \infty, \forall \tau \in \mathbb{R}^d$. Then*

$$I(\beta) = \sup_{\tau \in \mathbb{R}^d} [\langle \tau, \beta \rangle - H(\tau)],$$

where

$$H(\tau) = \lim_{k\to\infty} \frac{1}{k} \log E e^{k\langle \tau, X_k \rangle}.$$

With Proposition 5.3, we are ready to prove Theorem 5.1.

Proof of Theorem 5.1. We carry out the proof by adopting and modifying the proof of [14, Theorem 6.4.4,p.279]. Choose a concave continuous function $\zeta : \mathbb{R}^d \to [-B, 0]$. Since X_k takes values on a compact subset $K_c \subset \mathbb{R}^d$, ζ is Lipschitz continuous on K_c. Assume $|\zeta(x) - \zeta(y)| < G|x-y|$ for all $x, y \in K$. Set $C = \sup_{x\in K_c} \zeta(x)$ and

$$\widehat{S}_k^m = \frac{X_{m+1} + \cdots + X_k}{k-m}.$$

Observe that

$$\left| \widehat{S}_{k+m} - \left(\frac{k}{k+m} \widehat{S}_k + \frac{m}{k+m} \widehat{S}_{k+m+l}^{k+l} \right) \right| \leq \frac{2lC}{k+m}.$$

By Lipschitz continuity of ζ,

$$\left|\zeta(\widehat{S}_{k+m}) - \zeta\left(\frac{k}{k+m}\widehat{S}_{m_0} + \frac{m}{k+m}\widehat{S}^{k+l}_{k+m+l}\right)\right| \leq \frac{2lCG}{k+m}.$$

Since ζ is concave,

$$(m+k)\zeta(\widehat{S}_{m+k}) \geq k\zeta(\widehat{S}_k) + m\zeta(\widehat{S}^{k+l}_{k+m+l}) - 2lCG. \tag{5.2}$$

Set

$$W = \mathrm{e}^{k\zeta(\widehat{S}_k)}, \quad Z = \mathrm{e}^{m\zeta(\widehat{S}^{k+l}_{k+m+l})}.$$

By virtue of assumption (A5.1),

$$EWEZ - EWZ \leq \mathrm{e}^{-r(l)l}. \tag{5.3}$$

Noting that $EWEZ \geq -(m+k)B$ and letting $l = m+k$, from Eq. (5.3),

$$\log EWZ \geq \log EWEZ + \log(\mathrm{e}^{(B-r(m+k))(m+k)}). \tag{5.4}$$

Since $r(k+m) \to \infty$ as $m, k \to \infty$, we can choose m, k large enough such that $\log EWZ \geq \log EWEZ + \frac{1}{2}$. Let $h(k) = \log E\mathrm{e}^{k\zeta(\widehat{S}_k)}$. By the above inequality and Eq. (5.2), we have

$$h(m+k) \leq h(m) + h(k) + \frac{1}{2} + 2r(m+k)CG. \tag{5.5}$$

Applying assumption (A5.1) yields that

$$H_g = \lim_{k\to\infty} \frac{1}{k} \log E\mathrm{e}^{k\zeta(X_k)}$$

exists, and hence by Proposition 5.3(a), $\{\widehat{S}_k\}$ satisfies the LDP with the rate function

$$I(\beta) = \sup_{f \in C_b(\mathbb{R}^d)} [f(\beta) - H_f].$$

Next, we need to show that $I(\cdot)$ is convex to reduce the form of $I(\cdot)$ to Eq. (5.1) by means of Proposition 5.3(c). To show that $I(\cdot)$ is convex, we need only show that if for some $M < \infty$ and fixed open sets G and $\widetilde{G}$.

$$P(\widehat{S}_k \in G)P(\widehat{S}_k \in \widetilde{G}) \geq \exp(-Mk)$$

for all k large enough, then

$$\liminf_{\eta\downarrow 0} \liminf_{k\to\infty} \rho(k,\eta) \geq 0, \tag{5.6}$$

where

$$\rho(k,\eta) = \frac{1}{k} \log \frac{P(\widehat{S}_k \in G, \widehat{S}^{k+\eta k}_{2k+\eta k} \in \widetilde{G})}{P(\widehat{S}_k \in G)P(\widehat{S}^{k+\eta k}_{2k+\eta k} \in \widetilde{G})}.$$

Using assumption (A5.1) again, we obtain

$$\begin{aligned}&\frac{P(\widehat{S}_k \in G, \widehat{S}_{2k+\eta k}^{k+\eta k} \in \widetilde{G})}{P(\widehat{S}_k \in G)P(\widehat{S}_{2k+\eta k}^{k+\eta k} \in \widetilde{G})}\\&\geq 1 - \frac{\mathrm{e}^{-r(k\eta)(k\eta)}}{P(\widehat{S}_k \in G)P(\widehat{S}_{2k+\eta k}^{k+\eta k} \in \widetilde{G})}\\&\geq 1 - \mathrm{e}^{(M-r(k\eta)k\eta)}.\end{aligned} \tag{5.7}$$

Thus

$$\begin{aligned}&\liminf_{\eta\downarrow 0}\liminf_{k\to\infty}\rho(k,\eta)\\&\geq \liminf_{\eta\downarrow 0}\liminf_{k\to\infty}\frac{1}{k}\log(1-\mathrm{e}^{(M-r(k\eta)k\eta)})\\&=0.\end{aligned} \tag{5.8}$$

The proof of the theorem is complete. □

5.2 LDP for System Identification with Regular Sensors under Mixing Noises

We Apply Theorem 5.1 directly to the identification with regular sensors.

Theorem 5.4 *Assume that the noise sequence $\{d(l)\}$ satisfies assumption* (A5.1). *Then $\{\eta_k^s\}$ satisfies the LDP with the rate function*

$$\widetilde{I}(\widetilde{\beta}) = I(G^{-1}(\widetilde{\beta})) = I(\Phi_{m_0}(t_0)\widetilde{\beta})),$$

where

$$\begin{aligned}I(\beta) &= \sup_{\tau\in\mathbb{R}^d}[\langle\tau,\beta\rangle - H(\tau)],\\H(\tau) &= \lim_{k\to\infty}\log E\mathrm{e}^{k\langle\tau,X_k\rangle}.\end{aligned} \tag{5.9}$$

Proof. Let

$$Y_i = (d(t_0+(i-1)m_0+1),\ldots,d(t_0+(i-1)m_0+m_0))'.$$

Then $X_k = (Y_1+\cdots+Y_k)/k$. Since $\{d_i\}$ satisfies assumption (A5.1), so does $\{Y_i\}$. By Theorem 5.1, $\{X_k\}$ satisfies the LDP with rate function I. Then by the contraction principle, Proposition 3.5, $\{\eta_k^s\}$ satisfies the LDP with the rate function

$$\widetilde{I}(\widetilde{\beta}) = I(G^{-1}(\widetilde{\beta})) = I(\Phi_0\widetilde{\beta}).$$

The proof is complete. □

5.3 LDP for Identification with Binary Sensors under Mixing Conditions

Recall the notation in Chap. 2 and apply Theorem 5.1.

Theorem 5.5 *Assume that the noise sequence* $\{d(l)\}$ *satisfies the first part of assumption* (A5.1). *Then* $\{\widehat{\theta}_k\}$ *satisfies the LDP with the rate function* $\widehat{I}(\widehat{\beta}) = I(F(C - \Phi_0\widehat{\beta}))$, *where*

$$\begin{aligned} I(\beta) &= \sup_{\tau\in\mathbb{R}^d}[\langle\tau,\beta\rangle - H(\tau)], \\ H(\tau) &= \lim_{k\to\infty} \log E e^{k\langle\tau,X_k\rangle}. \end{aligned} \tag{5.10}$$

Proof. Let

$$Y_i = (\chi_{\{d(t_0+(i-1)m_0+1)\}}, \dots, \chi_{\{d(t_0+(i-1)m_0+m_0)\}})'.$$

Then

$$X_k = (Y_1 + \cdots + Y_k)/k.$$

Since $\{d(l)\}$ satisfies the first part of assumption (A5.1) and $|Y_i| \le m_0$, $\{Y_i\}$ satisfies assumption (A5.1). By Theorem 5.1 and the fact that $\{X_k\}$ satisfies the LDP with rate function I and by virtue of the contraction principle, Proposition 3.5, $\{\widehat{\theta}_k\}$ satisfies the LDP with rate function $\widehat{I}(\widehat{\beta}) = I(F(C - \Phi_0\widehat{\beta}))$. The proof is concluded. □

6
Applications to Battery Diagnosis

This chapter uses a battery diagnosis problem to illustrate the use of the LDP in industrial applications. Management of battery systems plays a pivotal role in electric and hybrid vehicles, and in support of distributed renewable energy generation and smart grids. The state of charge (SOC), the state of health (SOH), internal impedance, open circuit voltage, and other parameters indicate jointly the state of the battery system. Battery systems' behavior varies significantly among individual cells and demonstrates time-varying and nonlinear characteristics under varying operating conditions such as temperature, charging/discharging rates, and different SOCs. Such complications in capturing battery features point to a common conclusion that battery monitoring, diagnosis, and management should be individualized and adaptive. Although battery health diagnosis has drawn substantial research effort recently [26, 56] using various estimation methods or signal-processing algorithms [6, 43], the LDP has never been used before in this application. In the following, we describe a joint estimation algorithm for real-time battery parameter and SOC estimation that supports a diagnosis method using LDPs. For further reading on power control and battery systems, see [4, 5, 9, 13, 20, 21, 24, 25, 34, 44, 52] and references therein.

6.1 Battery Models

To illustrate, we use the battery model from the demonstration case in the SimPower Systems Toolbox in Matlab/Simulink (Mathworks [39, 53]),

Q. He et al., *System Identification Using Regular and Quantized Observations*, SpringerBriefs in Mathematics, DOI 10.1007/978-1-4614-6292-7_6,

which was used and simplified in [33] for battery model identification. Define v = terminal voltage (V), E_0 = internal open-circuit voltage (V), K = polarization constant (see the footnote on the next page for a remark on this parameter), i = battery current (A), q = extracted capacity (Ah), Q = maximum battery capacity (Ah), T = battery response time constant (second). Let s be the SOC and w the state variable for the battery dynamic delay time (a first-order system). The basic relationship between the SOC and q is

$$s(t) = \frac{Q - q(t)}{Q},$$

or

$$q(t) = Q(1 - s(t)). \tag{6.1}$$

Using a unit conversion, $q(t)$ may be calculated by integration. We have

$$\begin{aligned} q(t) &= q(0) + \int_0^t i(\tau)\mathrm{d}\tau/3600 \\ &= \left[\int_0^t i(\tau)\mathrm{d}\tau + 3600Q(1 - s(0))\right]/3600. \end{aligned}$$

This relationship, however, cannot be used alone in estimating the SOC, since $s(0)$ is unknown and the integration will create accumulated errors. In this model structure, the dynamic part of the battery model is linear [39]:

$$\begin{cases} \dot{s} = -\dfrac{1}{Q}\dot{q} = -\dfrac{1}{Q}\dfrac{i}{3600}, \\ \dot{w} = -\dfrac{1}{T}w + \dfrac{1}{T}i\,. \end{cases}$$

The output of the filter is $w(t)$, which differs from $i(t)$ only for a short period of time. Under a constant current, it is reasonable for it to equal $w(t)$ and $i(t)$ in the output equation after a small transient period. Define $x = [s, w]'$. We have

$$\dot{x} = Ax + Bi = \begin{bmatrix} 0 & 0 \\ 0 & -\dfrac{1}{T} \end{bmatrix} x + \begin{bmatrix} -\dfrac{1}{3600Q} \\ \dfrac{1}{T} \end{bmatrix} i. \tag{6.2}$$

The output is the terminal voltage v, which is a nonlinear function of s,w, and i. The function differs under charge or discharge operations. Under discharge operation,

$$v = f_d(x, i) = E_0 + ae^{-bQ(1-s)} - K\frac{Q(1-s)}{s} - K\frac{w}{s} - iR, \tag{6.3}$$

and under charge operation,

$$v = f_c(x, i) = E_0 + ae^{-bQ(1-s)} - K\frac{Q(1-s)}{s} - K\frac{w}{1.1 - s} - iR. \tag{6.4}$$

In practical settings, i and v are subject to measurement noises

$$u = i + n, \quad y = v + d, \tag{6.5}$$

where n and d are noises whose properties will be specified later.

The discharge model (6.3) contains several terms

$$v = E_0 + ae^{-bQ(1-s)} - K\frac{Q(1-s)}{s} - K\frac{w}{s} - iR.$$

During battery operations, the cell is always working in the normal range, typically with the SOC in the range $0.4 \le s \le 0.8$. Consequently, the terms that represent features outside this range may be omitted in the first approximation. The exponential term $ae^{-bQ(1-s)}$ is nearly zero for $s \le 0.8$ and can be omitted. The term $Kw(\frac{1}{s})$ contains the output of the filter, which has a much faster dynamics than the main storage dynamics. Consequently, after an initial transient period, we may replace w by i. These steps lead to a simplified model

$$v = E_0 - K\frac{Q(1-s)}{s} - K\frac{i}{s} - Ri, \tag{6.6}$$

which contains the four unknown parameters E_0, K, Q, R.

6.2 Joint Estimation of Model Parameters and SOC

The main ideas of joint estimation of SOC and model parameters were introduced in [33]. Suppose that the starting time for discharge is t_0 and the SOC at t_0 is $s(t_0) = s_0$, which is unknown. From $\dot{s} = -\frac{1}{Q}\frac{i}{3600}$ we have $s(t) = s_0 - \frac{\lambda(t)}{Q}$, where $\lambda(t) = \frac{1}{3600}\int_{t_0}^{t} i(\tau)\mathrm{d}\tau$, and $\lambda(t)$ is calculated from measured $i(t)$ and hence is known. The simplified output equation (6.6) is

$$\begin{aligned} v(t) &= E_0 - K\frac{Q(1-s(t))}{s(t)} - K\frac{i(t)}{s(t)} - Ri(t) \\ &= E_0 + KQ - \frac{KQ}{s_0 - \frac{1}{Q}\lambda(t)} - \left(\frac{K}{s_0 - \frac{1}{Q}\lambda(t)} + R\right) i \\ &= h(t) - g(t)i(t), \end{aligned}$$

where

$$h(t) = E_0 + KQ - \frac{KQ}{s_0 - \frac{1}{Q}\lambda(t)}, \quad g(t) = \frac{K}{s_0 - \frac{1}{Q}\lambda(t)} + R.$$

Note that $h(t)$ and $g(t)$ contain $\lambda(t)$, which is an integral function, and hence slowly time-varying. Suppose that in a small time interval $(t-\tau, t]$, we add an i.i.d. dither sequence $\{\psi_k : k = 1, \ldots, N\}$ of mean zero and variance σ^2, namely $E\varepsilon_1 = 0$ and $E\varepsilon_1^2 = \sigma^2$. Set $\tau_k = t - \tau + \frac{\tau}{N}k$, $k = 1, \ldots, N$. Under a smooth current load $i(t)$,

$$v(\tau_k) = h(\tau_k) - g(\tau_k)(i(\tau_k) + \psi_k), \quad k = 1, \ldots, N. \tag{6.7}$$

For sufficiently small τ,

$$h(\tau_k) \approx h(t), \quad g(\tau_k) \approx g(t), \quad i(\tau_k) \approx i(t),$$

since $\lambda(t)$ and $i(t)$ are smooth. As a result, Eq. (6.7) may be written as

$$v(\tau_k) = h(t) - g(t)(i(t) + \psi_k), \quad k = 1, \ldots, N. \tag{6.8}$$

The above process can be repeated at different t. Suppose that we have identified values of $h(t)$ and $g(t)$ at t_j, $j = 1, \ldots, m$. Namely, we have the equations

$$E_0 + KQ - \frac{KQ}{s_0 - \frac{1}{Q}\lambda(t_j)} = h(t_j), \quad j = 1, \ldots, m,$$
$$\frac{K}{s_0 - \frac{1}{Q}\lambda(t_j)} + R = g(t_j), \quad j = 1, \ldots, m.$$

The remaining task is to calculate the five unknown constants R, K, E_0, Q, s_0 from these equations. Denote the true parameter by $\theta^* = [R, K, E_0, Q, s_0]'$ and their estimates at kth iteration step by $\widehat{\theta}_k$.

Define

$$s(t) = s_0 - \frac{1}{Q}\lambda(t).$$

Since

$$\begin{aligned}
g_1(t) &= \frac{\partial g}{\partial R} = 1, \\
g_2(t) &= \frac{\partial g}{\partial K} = \frac{1}{s(t)}, \\
g_3(t) &= \frac{\partial g}{\partial E_0} = 0, \\
g_4(t) &= \frac{\partial g}{\partial Q} = -\frac{K\lambda(t)}{Q^2(s(t))^2}, \\
g_5(t) &= \frac{\partial g}{\partial s_0} = -\frac{K}{(s(t))^2}, \\
h_1(t) &= \frac{\partial h}{\partial R} = 0, \\
h_2(t) &= \frac{\partial h}{\partial K} = Q\left(1 - \frac{1}{s(t)}\right), \\
h_3(t) &= \frac{\partial h}{\partial E_0} = 1, \\
h_4(t) &= \frac{\partial h}{\partial Q} = K + K\frac{s(t) - \frac{\lambda(t)}{Q}}{(s(t))^2}, \\
h_5(t) &= \frac{\partial h}{\partial s_0} = \frac{KQ}{(s(t))^2},
\end{aligned}$$

the Jacobian matrix can be numerically evaluated as

$$J(\theta) = \begin{bmatrix} g_1(t_1) & g_2(t_1) & g_3(t_1) & g_4(t_1) & g_5(t_1) \\ \vdots & & & & \vdots \\ g_1(t_m) & g_2(t_m) & g_3(t_m) & g_4(t_m) & g_5(t_m) \\ h_1(t_1) & h_2(t_1) & h_3(t_1) & h_4(t_1) & h_5(t_1) \\ \vdots & & & & \vdots \\ h_1(t_m) & h_2(t_m) & h_3(t_m) & h_4(t_m) & h_5(t_m) \end{bmatrix}.$$

Starting from an initial estimate $\widehat{\theta}_0$, the iterative search algorithm is the gradient-based search

$$\widehat{\theta}_{k+1} = \widehat{\theta}_k + \varepsilon(J(\widehat{\theta}_k)'J(\widehat{\theta}_k))^{-1}J(\widehat{\theta}_k)'(H - F(\widehat{\theta}_k) + \delta_k), \qquad (6.9)$$

where δ_k is the disturbance on H due to estimation of g and h.

6.3 Convergence

Suppose that $\{\delta_k\}$ in Eq. (6.9) is a sequence of (vector-valued) i.i.d. (independent and identically distributed) random variables with $E\delta_k = 0$ and $E\delta_k\delta_k' = \Sigma$. Convergence analysis of the recursive algorithm (6.9) can be conducted by the ODE (ordinary differential equation) approach [31]. Define a piecewise constant interpolation sequence $\widehat{\theta}^\varepsilon(t) = \widehat{\theta}_k$ for $t \in [\varepsilon k, \varepsilon k + \varepsilon)$. Then using martingale averaging methods in [31, Chap. 8], it can be shown that as $\varepsilon \to 0$, $\widehat{\theta}^\varepsilon(\cdot)$ converges weakly to $\theta(\cdot)$ such that $\theta(\cdot)$ is the solution of the ODE

$$\dot{\theta} = M(\theta)(H - F(\theta)), \qquad (6.10)$$

where

$$M(\theta) = (J(\theta)'J(\theta))^{-1}J(\theta).$$

Since $H - F(\theta^*) = 0$, the true parameter θ^* is an equilibrium point of Eq. (6.10). Furthermore, the Jacobian matrix at θ^*,

$$\Phi = -(J(\theta)'J(\theta))^{-1}J(\theta)\frac{\partial F(\theta)}{\partial \theta}, \qquad (6.11)$$

is negative definite in a neighborhood of θ^*.

To study the quality of approximation, namely, rate of convergence, we further consider the centered and scaled estimation errors

$$z_k = (\widehat{\theta}_k - \theta^*)/\sqrt{\varepsilon}.$$

It can be shown as in [31, Chap. 10] that for sufficiently large N_ε, the sequence $\{z_k : k \geq N_\varepsilon\}$ is tight (i.e., no probability is lost). Define the interpolated process

$$z^\varepsilon(t) = z_k \quad \text{for} \quad t \in [\varepsilon(k - N_\varepsilon), \varepsilon(k - N_\varepsilon + 1)).$$

Then we can further show that $z^\varepsilon(\cdot)$ converges weakly to $z(\cdot)$, which is a solution of the stochastic differential equation

$$\mathrm{d}z = \Phi z \mathrm{d}t + \Gamma \mathrm{d}w, \tag{6.12}$$

where Φ is defined in Eq. (6.11),

$$\Gamma = (J(\theta)' J(\theta))^{-1} J(\theta) \Sigma^{1/2},$$

and w is a standard Wiener process. By solving the related Liapunov equation for Σ_0,

$$\Sigma_0 \Phi' + \Phi \Sigma_0 = -\Gamma \Gamma',$$

we can find the stationary covariance (or asymptotic covariance) Σ_0. This yields that $z^\varepsilon(\cdot + \widehat{T})$ converges to a stationary Gauss Markov process as $\varepsilon \to 0$ and $\widehat{T} \to \infty$. The rate of convergence is determined by the scaling factor $\sqrt{\varepsilon}$ together with the asymptotic covariance Σ_0. Thus roughly,

$$\widehat{\theta}_k - \theta^* \sim N(0, \varepsilon \Sigma_0), \quad \text{for sufficiently small} \quad \varepsilon \quad \text{and large} \quad k. \tag{6.13}$$

6.4 Probabilistic Description of Estimation Errors and Diagnosis Reliability

The asymptotic normality in Eq. (6.13) indicates that for sufficiently large k and small ε, z_k is approximately normally distributed. Consequently, the estimate $\widehat{\theta}_k$ is also approximately normally distributed with

$$\widehat{\theta}_k \sim N(\theta^*, \varepsilon \Sigma_0).$$

Note that if we assume that $\{\delta_k\}$ is a sequence of i.i.d. random variables with Gaussian distribution, then the distribution of $\widehat{\theta}_k$ is precisely Gaussian.

Diagnosis of battery conditions often involves one particular parameter at a time. For example, to evaluate the SOH, one uses the estimate $\widehat{Q}_k$ on Q. In the following derivations, we use the first component $\widehat{\theta}_k(1)$ as a concrete case for discussion.

Decompose Σ_0 into

$$\Sigma_0 = \begin{bmatrix} \sigma_{11}^2 & \Sigma_{12} \\ \Sigma_{12}' & \Sigma_{22} \end{bmatrix}.$$

By the well-known Slutsky theorem, for any constant vector v, $v'\widehat{\theta}_k$ is approximately normally distributed with mean $v'\theta^*$ and variance $v'\Sigma_0 v$. Thus, in particular, the marginal distribution of $\widehat{\theta}_k(1)$ is

$$\widehat{\theta}_k(1) \sim N(\theta^*(1), \varepsilon\sigma_{11}^2), \tag{6.14}$$

whose probability density function is given by

$$f_\varepsilon(x;\theta^*) = \frac{1}{\sqrt{\varepsilon}\sigma_{11}\sqrt{2\pi}} \mathrm{e}^{-\frac{(x-\theta^*(1))^2}{2\varepsilon\sigma_{11}^2}}. \tag{6.15}$$

In battery diagnosis, the "normal region" is defined by a set Ω_n, and the faulty region is designated by another set Ω_f, which is disjoint from Ω_n. The transitional region is $\Omega_t = \Omega - (\Omega_n \cup \Omega_f)$. If θ^* is outside Ω_n, the battery is considered "not normal" for that specific diagnosis. If $\theta^* \in \Omega_f$, the battery is considered "faulty." For diagnosis, we need their probabilistic descriptions.

From Eq. (6.14), if the true parameter is θ^*, then $\widehat{\theta}_k$ is approximately normally distributed with density function $f_\varepsilon(x;\theta^*)$. The diagnosis function $\mathcal{D}$ is defined as

$$\mathcal{D}(\widehat{\theta}_k) = \begin{cases} \text{normal}, & \widehat{\theta}_k \in \Omega_n, \\ \text{warning}, & \widehat{\theta}_k \in \Omega_t, \\ \text{faulty}, & \widehat{\theta}_k \in \Omega_f. \end{cases}$$

Reliability of diagnosis is characterized by the following two error diagnoses:

Missed diagnosis: If $\theta^* \in \Omega_f$ but $\widehat{\theta}_k \in \Omega_n$, a missed diagnosis occurs when a faulty battery is diagnosed as normal. The probability of this diagnosis error is

$$\begin{aligned} P_{md} &= \sup_{\theta^* \in \Omega_f} P\{\mathcal{D}(\widehat{\theta}_k) = \text{"normal"} | \theta^*\} \\ &= \sup_{\theta^* \in \Omega_f} P\{\widehat{\theta}_k \in \Omega_n | \theta^*\} \\ &= \sup_{\theta^* \in \Omega_f} \int_{\Omega_n} f_\varepsilon(x;\theta^*)\mathrm{d}x. \end{aligned}$$

False diagnosis: If $\theta^* \in \Omega_n$ but $\widehat{\theta}_k \in \Omega_f$, a false diagnosis occurs if a normal battery is diagnosed as faulty. The probability of this diagnosis error is

$$\begin{aligned} P_{fd} &= \sup_{\theta^* \in \Omega_n} P\{\mathcal{D}(\widehat{\theta}_k) = \text{"faulty"} | \theta^*\} \\ &= \sup_{\theta^* \in \Omega_n} P\{\widehat{\theta}_k \in \Omega_f | \theta^*\} \\ &= \sup_{\theta^* \in \Omega_n} \int_{\Omega_f} f_\varepsilon(x;\theta^*)\mathrm{d}x. \end{aligned}$$

6.5 Computation of Diagnosis Reliability

Consider the case of battery capacity Q (Ah). Suppose that the new battery capacity is $Q_{\max}$. Then, $\Omega = [0, Q_{\max}]$. For vehicle applications, the battery can provide rated performance if $Q \in [\alpha Q_{\max}, Q_{\max}]$ for some $0 \leq \alpha < 1$. A typical value of α is 75%. If $Q \leq \beta Q_{\max}$ for some $0 \leq \beta < \alpha$ (e.g., $\beta = 50\%$), then the battery must be retired. When $Q \in (\beta Q_{\max}, \alpha Q_{\max})$, a replacement recommendation is issued, but replacement is not mandatory. As a result, we have $\Omega_n = [\alpha Q_{\max}, Q_{\max}]$, $\Omega_t = (\beta Q_{\max}, \alpha Q_{\max})$, $\Omega_f = [0, \beta Q_{\max})$. These regions and relations to the density function of $\widehat{\theta}_k$ are shown in Fig. 6.1.

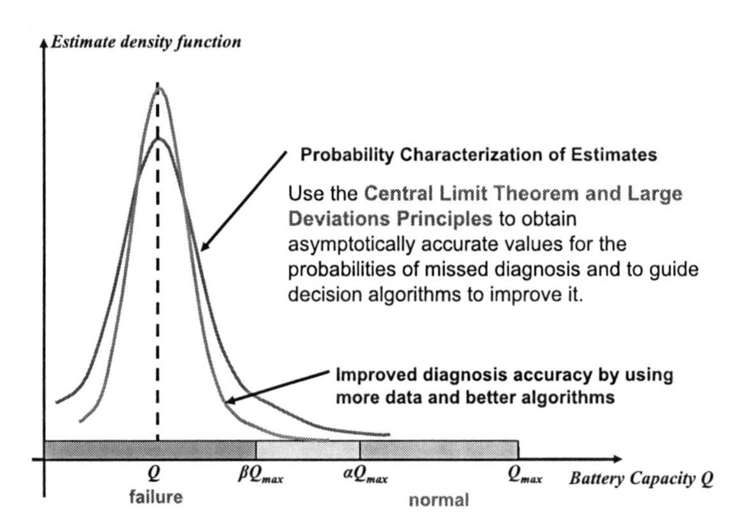

FIGURE 6.1. Estimated density function and diagnosis regions

For simplicity, use $\sigma = \sigma_{11}$ in Eq. (6.15) in the following theorem, whose proof is omitted.

Theorem 6.1

$$P_{md} = \frac{1}{c(\beta Q_{\max})\sqrt{\varepsilon}\sigma\sqrt{2\pi}} \int_{\alpha Q_{\max}}^{Q_{\max}} \mathrm{e}^{-\frac{(x-\beta Q_{\max})^2}{2\varepsilon\sigma^2}} \mathrm{d}x$$

and

$$P_{fd} = \frac{1}{c(\alpha Q_{\max})\sqrt{\varepsilon}\sigma\sqrt{2\pi}} \int_{0}^{\beta Q_{\max}} \mathrm{e}^{-\frac{(x-\alpha Q_{\max})^2}{2\varepsilon\sigma^2}} \mathrm{d}x,$$

where

$$c(Q) = \frac{1}{\sqrt{\varepsilon}\sigma\sqrt{2\pi}} \int_{0}^{Q_{\max}} \mathrm{e}^{-\frac{(x-Q)^2}{2\varepsilon\sigma^2}} \mathrm{d}x.$$

6.6 Diagnosis Reliability via the Large Deviations Principle

For small ε, P_{md} (and similarly for P_{fd}) can be expressed as follows. We shall include ε explicitly in $c_\varepsilon(Q)$ and $P_{md}(\varepsilon)$. Since $\varepsilon\sigma^2$ is the variance, it is easy to verify that

$$c_\varepsilon(Q) \to 1 \text{ and } P_{md}(\varepsilon) \to 0, \text{ as } \varepsilon \to 0.$$

First, we note that Taylor expansions will not work, since we can establish that for every n,

$$\left.\frac{\partial^n P_{md}(\varepsilon)}{\partial \varepsilon^n}\right|_{\varepsilon=0} = 0.$$

This is, in fact a conclusion from the large deviations principle (Theorem 3.1), which indicates that $P_{md}(\varepsilon)$ converges to 0 exponentially fast. Consequently, we employ LDP to derive asymptotic expressions for $P_{md}(\varepsilon)$.

For asymptotic analysis, we may ignore the truncation and use the standard normal density function $f(x; Q)$ in Eq. (6.15) directly. From Theorem 4.3 and Remark 4.4, for the normal distribution (6.15), the LDP states that

$$\varepsilon \ln P\{|\widehat{Q}_k - Q^*| \geq \delta\} \to -\frac{\delta^2}{2\sigma^2} \quad \text{as} \quad \varepsilon \to 0,$$

where $\frac{\delta^2}{2\sigma^2}$ is the rate function in Eq. (4.5). Or for small ε,

$$P\{|\widehat{Q}_k - Q^*| \geq \delta\} \approx \kappa(\varepsilon) \mathrm{e}^{-\frac{\delta^2}{2\varepsilon\sigma^2}} \tag{6.16}$$

so that $\varepsilon \ln \kappa(\varepsilon) \to 0$, $\varepsilon \to 0$. In general, the LDP focuses on the rate functions and does not consider further details on $\kappa(\varepsilon)$; see the generic constant κ in Eq. (1.1). In our special case, however, $\kappa(\varepsilon)$ can be further calculated, but its proof is omitted here.

Theorem 6.2

$$\sqrt{\varepsilon}\kappa(\varepsilon) \to \sqrt{\frac{2}{\pi}} \frac{\sigma}{\delta} \quad \textit{as} \quad \varepsilon \to 0, \tag{6.17}$$

and

$$P\{|\widehat{Q}_k - Q^*| \geq \delta\} \approx \sqrt{\frac{2}{\pi}} \frac{\sigma}{\sqrt{\varepsilon}\delta} \mathrm{e}^{-\frac{\delta^2}{2\varepsilon\sigma^2}}. \tag{6.18}$$

In particular, for P_{md} as in Theorem 6.1, we have

$$\begin{aligned} P_{md} &= P\{\widehat{Q}_k \in [0, \beta Q_{\max}] | \alpha Q_{\max}\} \\ &\leq P\{\widehat{Q}_k \in (-\infty, \beta Q_{\max}] | \alpha Q_{\max}\} \\ &\leq P\{|\widehat{Q}_k - \alpha Q_{\max}| \geq (\alpha - \beta) Q_{\max}\} \\ &\approx \sqrt{\frac{2}{\pi}} \frac{\sigma}{\sqrt{\varepsilon}\delta} \mathrm{e}^{-\frac{(\alpha-\beta)^2 Q_{\max}^2}{2\varepsilon\sigma^2}}. \end{aligned}$$

The expression

$$P(\varepsilon) = \sqrt{\frac{2}{\pi}} \frac{\sigma}{\sqrt{\varepsilon}\delta} e^{-\frac{(\alpha-\beta)^2 Q_{\max}^2}{2\varepsilon\sigma^2}} \tag{6.19}$$

can be used in diagnosis reliability analysis. Similarly, we can show that

$$P_{fd} \leq P(\varepsilon).$$

Since both P_{md} and P_{fd} share the same upper bound $P(\varepsilon)$, we will refer to $P(\varepsilon)$ as the reliability measure.

Example 6.3 The Li-ion battery has a rated capacity of 10 (Ah). The battery is considered healthy if its capacity is within 80% of its rated value, namely $Q \geq 8$ (Ah). When the battery capacity value falls below 70% of its rated capacity, i.e., 7 (Ah), it is to be retired from automotive use. Our diagnosis algorithm employs the joint identification algorithm to estimate the model parameters and SOC together. For capacity diagnosis (i.e., SOH evaluation), we will use only the estimates on Q. Suppose that the true capacity of the battery has decreased to 6.7 (Ah). The true parameters of the battery are $R = 0.3077, K = 0.1737, E_0 = 216.6753, Q = 6.7$. The actual initial SOC is $s_0 = 0.5$, which is to be estimated also. The values of $g(t)$ and $h(t)$ at the selected discharge steps are estimated using dithered input. From an initial SOC s_0, estimation of $g(t)$ and $h(t)$ is performed at a discharge depth from $1 - s_0 + 0.01$ (SOC $= s_0 - 0.01$) to $1 - s_0 + 0.2$ (SOC $= s_0 - 0.2$) with 0.01 increment. So, at each selected discharge depth, in a very small time interval, a dither of i.i.d. random variables of length $N = 2000$ is added to the load current, which is a constant $i_0 = 2$ (A). The dither sequence is uniformly distributed in $[-0.4, 0.4]$. As a result, it has mean zero and variance $\sigma^2 = 0.4^2/3 = 0.053$. From the measured terminal voltage $v(k)$ at these 2000 sampled points, we use the Matlab functions $\mu = \text{mean}(v)$ and $\widehat{\sigma}^2 = \text{var}(v)$ to calculate its sample mean and sample variance. Then, the estimated function values are derived from

$$\widehat{g} = \sqrt{\widehat{\sigma}^2/0.053}, \quad \widehat{h} = \mu + \widehat{g} i_0.$$

This process is repeated over the selected points of discharge depths. These function values are then used in the above search algorithm to estimate the unknown model parameters R, K, E_0, Q and the unknown initial value s_0. The estimated value of Q at the exit is recorded.

The diagnosis decision function is

$$\mathcal{D}(\widehat{Q}) = \begin{cases} \text{normal}, & \widehat{Q} \geq 8l, \\ \text{warning}, & 7 < \widehat{Q} < 8l, \\ \text{faulty}, & \widehat{Q} \leq 7. \end{cases}$$

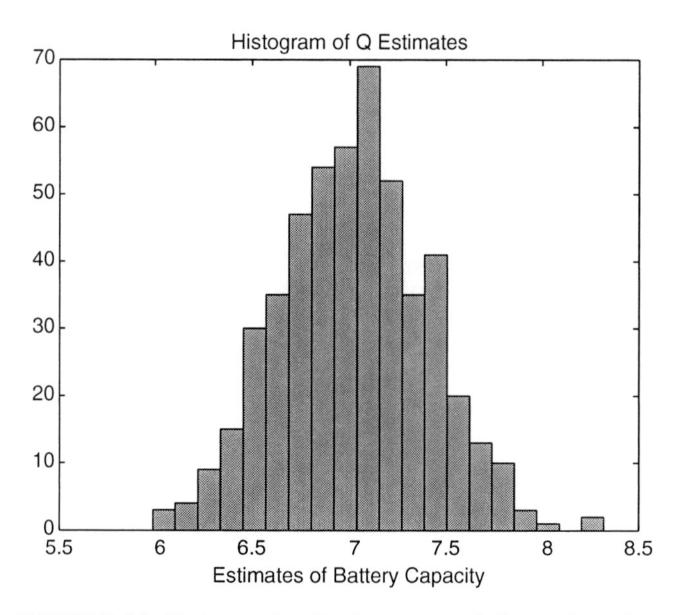

FIGURE 6.2. Estimate density function and diagnosis regions

To evaluate the diagnosis accuracy, the estimation scheme and diagnosis decision function are repeated 500 times. The resulting sample statistics of the Q estimates are as follows: The mean of the estimates is 7.0265, and the variance is 0.1425. The sample probability of missed diagnosis counts for the percentage of the Q estimates that fall into the normal range $Q \geq 8$ when the actual Q value is $Q \leq 7$. In this example, the actual Q value is 6.7. In this case study, among 500 samples, three were in the normal range, which amounts to 0.6% of diagnosis errors. The histogram of the estimates is plotted in Fig. 6.2.

7
Applications to Medical Signal Processing

Heart and lung sounds are of essential importance in medical diagnosis of patients with lung or heart diseases. To obtain reliable diagnosis and detection, it is critically important that cardiac and respiratory auscultation obtain sounds of high clarity. However, heart and lung sounds interfere with each other in auscultation, corrupting sound quality and causing difficulties in diagnosis. For example, the main frequency components of heart sounds, which are in the range of 50–100 Hz, often produce an intrusive interference that masks the clinical interpretation of lung sounds over the low-frequency band. It is highly desirable, especially in computerized heart/lung sound analysis, to separate the overlapped heart and lung sounds before using them for diagnosis.

This issue becomes further complicated when auscultation must be performed in operating rooms or other clinical environments. Unlike acoustic labs, in which noise levels can be artificially controlled and reduced, operating rooms are very noisy due to surgical devices, ventilation machines, conversations, alarms, etc. The unpredictable and broadband nature of such noises make operating rooms a very difficult acoustic environment. More technically, lung and heart sounds have frequency bands that overlap significantly with noise frequencies. As a result, high-fidelity microphones and traditional noise filtering or cancellation techniques cannot remove noises satisfactorily to obtain authentic sound signals.

From the viewpoint of signal processing, the above problem can be formulated as an adaptive blind source separation (BSS) problem, since the source signals cannot be measured directly and signal transmission channels

Q. He et al., *System Identification Using Regular and Quantized Observations*,
SpringerBriefs in Mathematics, DOI 10.1007/978-1-4614-6292-7_7,

are unknown. BSS is a very difficult problem, and it has drawn wide attention in various fields in recent years [23]. Different approaches have been developed such as output decorrelation, high-order statistics, and neural network methods. Most methods employ certain features that separate source signals such as different frequency bands (frequency-domain filtering techniques), signal stochastic independence (whitening filters), and statistics separation (statistics parameter extractions). For signals and noises whose separation of the above features is not significantly justified in applications, these methods become ineffective. Furthermore, it is well known that adaptive blind identification algorithms are often subject to slow convergence and high computational complexity.

This chapter is based on [73], which developed techniques for extracting authentic heart and lung sounds when auscultated sounds contain interference and noise corruption. Our approach utilizes the unique cyclic nature of respiratory and heart sounds to conduct channel identification, signal separation, and noise cancellation iteratively. The algorithm reconfigures the signal transmission channels during different phases of breathing and heartbeat cycles, translating a difficult blind adaptive noise cancellation (ANC) problem into a sequence of regular identification and noise cancellation problems. The reduced problems are easier to process and allow faster convergence rates and less computational burden. This technique does not require signal and noise to possess such separating features as frequency separation, stochastic independence, or distinctive parameters. As a result, it is more generic and can be applied to a broader spectrum of application areas as long as the source signals asynchronously go through existence and almost nonexistence stages cyclicly. We illustrate how the large deviations method can be used to treat such problems.

7.1 Signal Separation and Noise Cancellation Problems

We are considering two typical signal separation and noise cancellation problems. The first one is the signal separation problem shown in Fig. 7.1. The system contains two source signals s_1 and s_2. The measurements x_1 and x_2 are subject to cross interference from both source signals. A motivating example in our applications is separation of heart and lung sounds. The signal transmission channels are assumed to be linear time-invariant systems whose transfer functions are unknown. The goal is to generate authentic signals p_1 and p_2 using only the measurements x_1 and x_2. This is a problem of blind adaptive signal separation.

The key feature we use for signal separation is the cyclic nature of these two signals: Each signal undergoes the phases of emerging (inhalation and exhalation for lung sounds and heartbeat for heart sound) and pausing

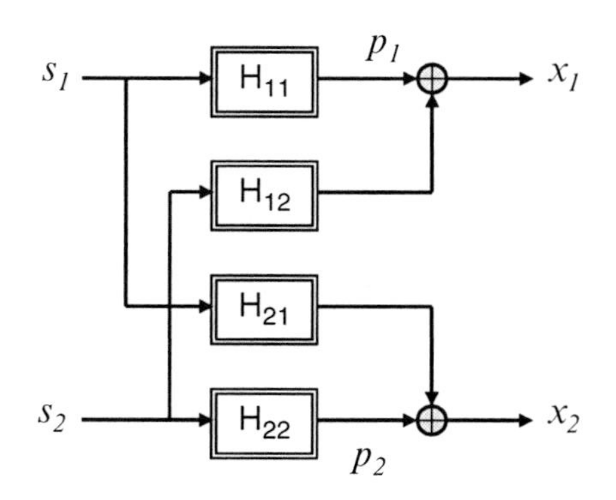

FIGURE 7.1. Two-signal source separation problems

(lung sound pausing between inhalation and exhalation and heart sound pausing in between heartbeats). It will become clear that this feature can be used effectively in separating the signals.

When the above system is further corrupted by noise, as shown in Fig. 7.2, the system becomes a three-signal separation problem. The main difference is that the two source signals s_1 and s_2 are assumed to possess the above cyclic nature, but the noise does not have any usable patterns. The noise may have spectra that overlap with those of the signals, and may be statistically correlated with signals. Due to different features in signals and noise, this problem will be treated as a combined problem of noise cancellation and signal separation.

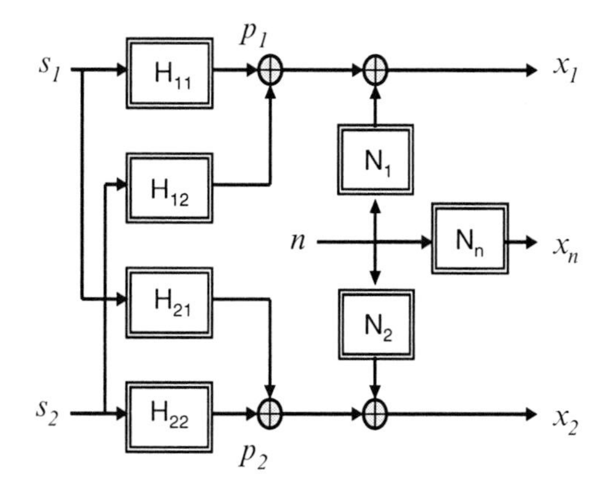

FIGURE 7.2. Source separation and noise cancellation problems

7.2 Cyclic System Reconfiguration for Source Separation and Noise Cancellation

The main approach of cyclic system reconfiguration will be explained in this section. We start with the scenario of signal separation. The main idea will be extended to more complicated problems of combined noise cancellation and signal separation.

7.2.1 Cyclic Adaptive Source Separation

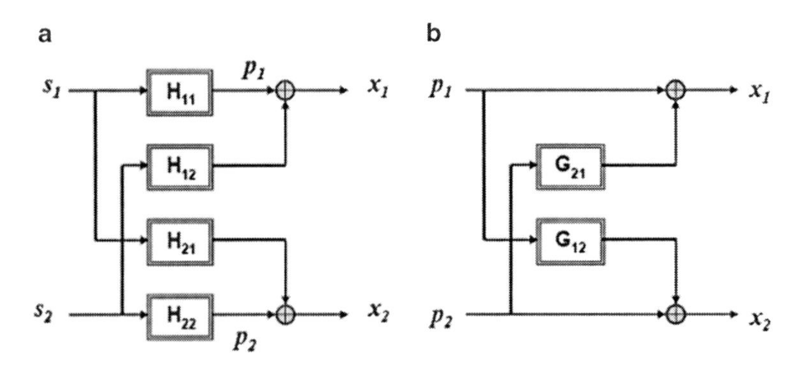

FIGURE 7.3. A fully coupled 2×2 system (part a) and its simplified representation (part b)

The 2×2 system (see Fig. 7.3a) has two signal sources, s_1, s_2, and two observations, x_1, x_2. The observations are related to the sources by the z-transfer function form

$$\begin{bmatrix} X_1(z) \\ X_2(z) \end{bmatrix} = \begin{bmatrix} H_{11}(z) & H_{12}(z) \\ H_{21}(z) & H_{22}(z) \end{bmatrix} \begin{bmatrix} S_1(z) \\ S_2(z) \end{bmatrix}$$

in Fig. 7.3a, which can be rewritten as

$$\begin{bmatrix} X_1(z) \\ X_2(z) \end{bmatrix} = \begin{bmatrix} 1 & G_{12}(z) \\ G_{21}(z) & 1 \end{bmatrix} \begin{bmatrix} P_1(z) \\ P_2(z) \end{bmatrix}$$

in Fig. 7.3b, where $P_1(z) = H_{11}(z)S_1(z)$, $P_2(z) = H_{22}(z)S_2(z)$, $G_{12}(z) = H_{12}(z)/H_{22}(z)$, $G_{21}(z) = H_{21}(z)/H_{11}(z)$; p_1 and p_2 are the signals we are looking for. Here p_1 and p_2 can be derived as

$$\begin{bmatrix} P_1(z) \\ P_2(z) \end{bmatrix} = \begin{bmatrix} 1 & G_{12}(z) \\ G_{21}(z) & 1 \end{bmatrix}^{-1} \begin{bmatrix} X_1(z) \\ X_2(z) \end{bmatrix}. \tag{7.1}$$

Since the signal transmission channels G_{12} and G_{21} are unknown, this is a typical BSS problem. There exist many approaches to the BSS problem such as output decorrelation, higher-order statistics, neural network based, minimum mutual information and maximum entropy, and geometric based. Although the underlying principles and approaches of those methods are different, most of these algorithms assume that the original signals are statistically independent.

Our method requires that the input signals undergo the intervals of emergence and near nonexistence sequentially. Many biomedical signals bear these features, such as heartbeats, lung sounds, and EEG signals. Our approach will use these features to cyclicly reconfigure the transmission channels so that the blind identification problem can be reduced into a number of regular identification problems.

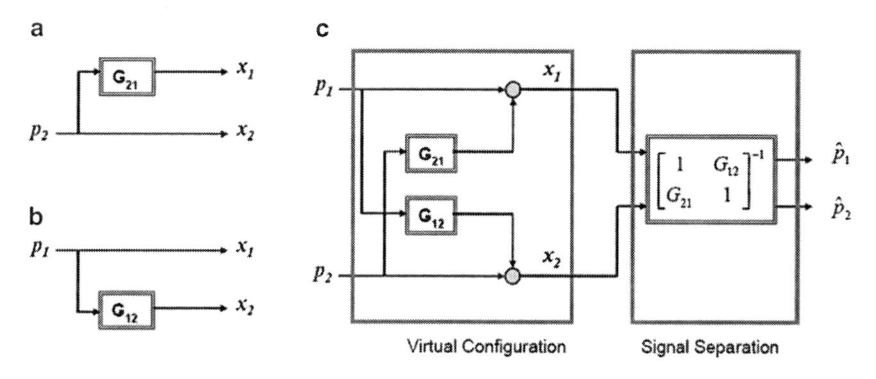

FIGURE 7.4. Cyclic reconfiguration of the system. (**a**) $p_1 = 0$: Identify G_{21}. (**b**) $p_2 = 0$: Identify G_{12}. (**c**) Signal separation

The following intervals are consequently recognized.

1) *Interval class I (shown in Fig. 7.4a)*: $p_1 \approx 0$ and p_2 is large. In this case, $X_1 = G_{2l}P_2$ and $X_2 = P_2$. As a result, sensor measurements x_1 and x_2 during interval class I can be used to identify the transmission channel G_{21}.

2) *Interval class II (shown in Fig. 7.4b)*: $p_2 \approx 0$ and p_1 is large. In this case, $X_2 = G_{12}P_1$ and $X_1 = P_1$. As a result, sensor measurements x_1, x_2 during interval class II can be used to identify the transmission channels G_{12}.

Once the transmission channels have been identified, we can obtain the estimates $\widehat{p}_1$ and $\widehat{p}_2$ using Eq. (7.1), shown in Fig. 7.4c.

7.2.2 Cyclic Adaptive Signal Separation and Noise Cancellation

The presence of noise in sensors is often inevitable. As shown in Fig. 7.5a, the 2×3 noisy system has two signal sources, s_1, s_2, and three sensors:

two signal sensors, x_1, x_2, and one noise reference sensor, x_n. The sensor observations are given by

$$\begin{bmatrix} X_1 \\ X_2 \end{bmatrix} = \begin{bmatrix} H_{11} & H_{12} \\ H_{21} & H_{22} \end{bmatrix} \begin{bmatrix} S_1 \\ S_2 \end{bmatrix} + \begin{bmatrix} N_1 \\ N_2 \end{bmatrix} \begin{bmatrix} N \end{bmatrix}.$$

Based on the virtual channel configuration in Fig. 7.5b, the equation can be rewritten as

$$\begin{bmatrix} X_1 \\ X_2 \end{bmatrix} = \begin{bmatrix} 1 & G_{21} \\ G_{12} & 1 \end{bmatrix} \begin{bmatrix} P_1 \\ P_2 \end{bmatrix} + \begin{bmatrix} N_{n1} \\ N_{n2} \end{bmatrix} \begin{bmatrix} X_n \end{bmatrix}, \tag{7.2}$$

where $G_{12} = H_{12}/H_{22}$, $G_{21} = H_{21}/H_{11}$, $N_{n1} = N_1/N_n$, $N_{n2} = N_2/N_n$.

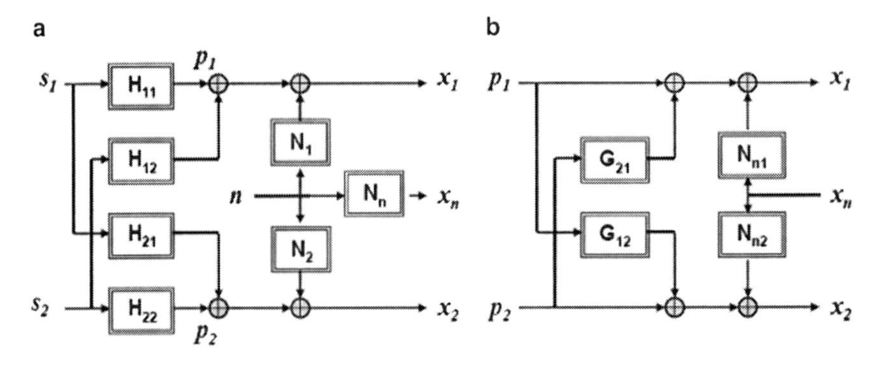

FIGURE 7.5. A noisy system (part a) and its simplified representation (part b)

To extract the desired signals p_1 and p_2, we first eliminate the noise, then perform the signal separation. Figure 7.6a–c show channel reconfigurations in different stages:

1. *Interval class I*: $p_1 \approx 0$ and $p_2 \approx 0$. In this case, Eq. (7.2) becomes

$$X_1 = N_{n1} X_n, \quad X_2 = N_{n2} X_n. \tag{7.3}$$

   As a result, sensor measurements x_1, x_2, and x_n during interval class I can be used to identify the noise transmission channels N_{n1} and N_{n2}.

2. *Interval class II*: $p_1 \approx 0$ and p_2 is large. In this case, Eq. (7.3) becomes $X_1 = G_{21} P_2 + N_{n1} X_n$, $X_2 = P_2 + N_{n2} X_n$, or

$$X_1 - N_{n1} X_n = G_{21}(X_2 - N_{n2} X_n). \tag{7.4}$$

   Since N_{n1} and N_{n2} have been identified during interval class I, this relationship can be used to identify G_{21}.

3. *Interval class III*: p_1 is large and $p_2 \approx 0$. In this case, Eq. (7.3) becomes $X_1 = P_1 + N_{n1}X_n$, $X_2 = G_{12}P_1 + N_{n2}X_n$, or

$$X_2 - N_{n2}X_n = G_{12}(X_1 - N_{n1}X_n). \tag{7.5}$$

   Since N_{n1} and N_{n2} have been identified during interval class I, this relationship can be used to identify G_{12}.

After channel identification, signals $\widehat{p}_1$ and $\widehat{p}_2$ are extracted as shown in Fig. 7.6d.

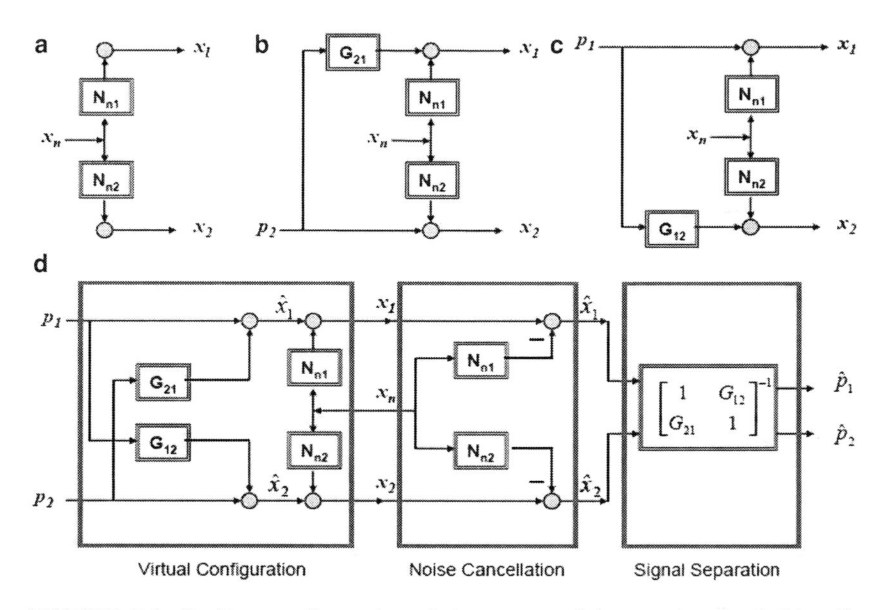

FIGURE 7.6. Cyclic reconfiguration of the system. (**a**) $p_1 = 0, p_2 = 0$: Identify N_{n1} and N_{n2}. (**b**) $p_1 = 0$: Identify G_{21}. (**c**) $p_2 = 0$: Identify G_{12}. (**d**) Noise cancellation and signal separation

7.3 Identification Algorithms

7.3.1 Recursive Time-Split Channel Identification

For identification of the virtual channels, we consider a generic input–output system $x = Gu + d$. Since G is stable, it can be modeled by its impulse response $g = \{g_0, g_1, \ldots\}$. Consequently, it can be represented in a regression form

$$x_k = \varphi_k' \theta + \widetilde{\varphi}_k' \widetilde{\theta} + d_k,$$

where

$$\varphi_k' = [u_k, \ldots, u_{k-n+1}]$$

is the principal regression vector, $\theta = [g_0, \ldots, g_{n-1}]'$ is the parameter vector of the modeled part of G, $\widetilde{\theta}_k' = [u_{k-n}, \ldots]$ and $\widetilde{\theta} = [g_n, \ldots]$ represent unmodeled dynamics, and d_k is disturbance. In this book, we will concentrate on the uncertainty from noises. The issue of unmodeled dynamics and its impact on identification accuracy was discussed in detail in [57]. There are several possible choices to identify G. We choose the standard least-squares estimation method for its relative simplicity and established convergence property [35]. Standard least-squares estimation leads to an estimate of the parameter θ of G, on the basis of N data points, as

$$\widehat{\theta}_N = \left(\frac{1}{N} \sum_{k=1}^{N} \varphi_k \varphi_k' \right)^{-1} \frac{1}{N} \sum_{k=1}^{N} \varphi_k x_k.$$

This process can be easily made recursive to reduce computational burden at each time instant, leading to a recursive least-squares algorithm

$$\begin{aligned} K_N &= \frac{P_{N-1}\varphi_{N-1}}{1 + \varphi_{N-1}' P_{N-1} \varphi_{N-1}}, \\ P_N &= (I - K_{N-1}\varphi_{N-1}') P_{N-1}, \\ \widehat{\theta}_N &= \widehat{\theta}_{N-1} + K_N (x_N - \varphi_N' \widehat{\theta}_{N-1}). \end{aligned} \tag{7.6}$$

The detailed recursive procedure is described as follows.

Step 1. Noise channel identification. During the pause stage of the kth cycle of Source-1 ($k = 0, 1, 2, \ldots$), when there's no source 2, the measured x_1, x_2, and x_n are used to identify the noise channels N_{n1} and N_{n2}, respectively, according to Eq. (7.3), using Eq. (7.6). The estimated model is denoted by $\widehat{N}_{n1}^k$ and $\widehat{N}_{n2}^k$.

Step 2. Noise cancellation and source-2-to-source-1 channel identification. During the pause stage of the kth cycle of source 1, when source 2 exists, the estimated noise channel models $\widehat{N}_{n1}^k$ and $\widehat{N}_{n2}^k$ are used to extract noise-free source-1 and source-2 measurements via

$$\widehat{x}_1^k = x_1^k - \widehat{N}_{n1}^k x_n^k, \quad \widehat{x}_2^k = x_2^k - \widehat{N}_{n2}^k x_n^k. \tag{7.7}$$

The estimated $\widehat{x}_2$ and $\widehat{x}_1$ are used as the input/output pair to identify the source-2-to-source-1 channel G_{21} using Eq. (7.6). The estimated model is denoted by $\widehat{G}_{21}^k$.

Step 3. Noise cancellation and source-1-to-source-2 channel identification. During the emergence stage of the kth cycle of source 1, when there is no source 2, the estimated noise channel models $\widehat{N}_{n1}^k$ and $\widehat{N}_{n2}^k$

are used to extract noise-free source-1 and source-2 measurements via Eq. (7.7). The estimated $\widehat{x}_1$ and $\widehat{x}_2$ are used as the input/output pair to identify the source-1-to-source-2 channel G_{12} using Eq. (7.6). The estimated model is denoted by $\widehat{G}_{12}^k$.

Step 4. Source-1 and source-2 separation. During the emergence stage of the kth cycle when both source 1 and source 2 exist, the estimated noise channel models $\widehat{N}_{n1}^k$ and $\widehat{N}_{n2}^k$ are used to extract noise-free source-1 and source-2 measurements via Eq. (7.7). The estimated source-1-to-source-2 and source-2-to-source-1 channel models $\widehat{G}_{12}^k$ and $\widehat{G}_{21}^k$ are used to separate source 1 and source 2 via

$$\begin{aligned} \widehat{p}_k &= \frac{\widehat{x}_1^k - \widehat{G}_{21}^k \widehat{x}_2^k}{1 - \widehat{G}_{12}^k \widehat{G}_{21}^k}, \\ \widehat{p}_k &= \frac{\widehat{x}_2^k - \widehat{G}_{12}^k \widehat{x}_1^k}{1 - \widehat{G}_{12}^k \widehat{G}_{21}^k}. \end{aligned} \tag{7.8}$$

Recursive steps: In the $(k+1)$th cycle of source 1, repeat steps 1–4. All estimated channel models are updated using the new data from the measured x_1, x_2, and x_n. The channel models derived from the previous cycles are used as the initial conditions, and the models are updated by RLS estimation. The newly updated virtual channel models are used to separate source 1 and source 2 in the $(k+1)$th cycle of source 1. These steps are then repeated from cycle to cycle.

7.3.2 Inversion Problem and Optimal Model Matching

It is noted that during source-1 and source-2 signal separation [see Eq. (7.8)], there is an inverse in the algorithm: $Q = 1/(1 - G_{12}G_{21})$. Although G_{12} and G_{21} are stable channels in practice, Q may not be stable, which will result in an unstable filter. The following two points are related to this issue:

(1) If the cross interference is relatively small, then Q will be stable. More precisely, suppose that $\|G_{12}\|_{H^\infty} < 1$ and $\|G_{21}\|_{H^\infty} < 1$, where $\|\cdot\|_{H^\infty}$ is the H^∞ norm. Then, by the small gain theorem, $1-G_{12}G_{21}$ has a stable inverse.

(2) If the cross interference is large, resulting in a large transfer function $G_{12}G_{21}$, then $1/(1 - G_{12}G_{21})$ may become unstable. To find a feasible signal separation algorithm, we will seek an approximate stable inverse of $M = 1 - G_{12}G_{21}$ by solving the optimal model matching problem

$$\inf_{Q \text{ is stable}} \|1 - MQ\|_{H^\infty}.$$

This is an optimal H^∞ design problem.

This problem can be solved by Nevalinna–Pick interpolation algorithms for finite-dimensional systems, state-space methods, or by numerical solutions using Matlab functions. Due to space limitations, the details of this design process are omitted here.

7.4 Quality of Channel Identification

In this section, we study asymptotic properties including convergence and rates of convergence of the recursive time-split algorithms proposed in the last section. For comparison, the traditional ANC is addressed here. Some of the theoretical results can be found in [72] and will not be presented here in detail.

7.4.1 Estimation Error Analysis for ANC

The ANC problem was first introduced in [65]. It is a special case of a 2×2 source separation problem, as shown in Fig. 7.4, where $s_1 = d$ is source, $s_2 = u$ is noise. The source-to-noise channel satisfies $G_{12} \approx 0$, and the noise-to-source channel is given by $G_{21} = G$. Here $x_k = \varphi_k'\theta + d_k$, where $\varphi_k' = [u_k, \ldots, u_{k-n+1}]$ is the principal regression vector and $\theta = [g_0, \ldots, g_{n-1}]'$ is the parameter vector of the modeled part of G. The standard ANC is based on the following basic procedures:

1) *Identification*: Standard least-squares estimation leads to an estimate of the parameter θ of G, on the basis of N data points, as

$$\widehat{\theta}_N = \left(\frac{1}{N}\sum_{k=1}^{N}\varphi_k\varphi_k'\right)^{-1}\frac{1}{N}\sum_{k=1}^{N}\varphi_k x_k$$

when $\frac{1}{N}\sum_{k=1}^{N}\varphi_k\varphi_k'$ is nonsingular.

2) *Noise cancellation*: The reconstructed signal $\widehat{d}_N$ at time N will be $\widehat{d}_N = x_N - \varphi_N'\widehat{\theta}_N$.

Assume that the underlying probability space is $\{\Omega, \mathcal{F}, P\}$. ANC works well under certain conditions, as discussed below. Consider the following terms in the estimation errors:

$$\begin{aligned} p_N &= \frac{1}{N}\sum_{k=1}^{N}\varphi_k\varphi_k', \\ q_N &= \frac{1}{N}\sum_{k=1}^{N}\varphi_k\widetilde{\varphi}_k'\widetilde{\theta}, \\ r_N &= \frac{1}{N}\sum_{k=1}^{N}\varphi_k d_k. \end{aligned}$$

Typically, the following assumptions are made on u_k and d_k in ANC. Under (A8.1), ANC provides appealing properties as indicated in Theorem 7.1; see the proof in [72].

(A7.1) (1) No unmodeled dynamics, that is, $\widetilde{\theta} = 0$. This requires a complex model structure to capture all channel dynamics. (2) $\{s_k\}$ is a stationary ergodic sequence satisfying $Eu_k = 0$. Here u_k is bounded by β uniformly in k and in $\omega \in \Omega$. (3) $\sum_{k=1}^{N} \varphi_k \varphi_k'/N$ is nonsingular for all $N \geq n+1$ with probability one (w.p.1). (4) $\{d_k\}$ is a sequence of i.i.d. random variables with $Ed_k = 0$, $Ed_k^2 = \sigma^2 > 0$, and $\{d_k\}$ is independent of $\{u_k\}$.

Theorem 7.1 *Under* (A7.1), *w.p.1,*

$$\frac{1}{N}\sum_{k=1}^{N} \varphi_k \varphi_k' \to M,$$
$$\left(\frac{1}{N}\sum_{k=1}^{N} \varphi_k \varphi_k'\right)^{-1} \to M^{-1},$$
$$r_N = \frac{1}{N}\sum_{k=1}^{N} \varphi_k d_k \to 0,$$

as $N \to \infty$, *where* M *and* M^{-1} *are positive definite matrices.*

7.4.2 Signal/Noise Correlation and the Large Deviations Principle

Assumption (A7.1), however, is often violated in this application. The independence between u_k and d_k is not always a valid assumption. When u_k and d_k are statistically correlated, accuracy of channel identification under ANC will be significantly compromised. To understand the impact of signal/noise correlation on estimation accuracy, hence quality of noise cancellation, we will investigate behavior of the estimates when u_k and d_k are correlated. Large deviations principles will be used to evaluate performance of algorithms under different noise characterizations. Consequently, the LDP forms a basis to objectively compare algorithms with respect to their estimation accuracy.

Assume that u_k and d_k are stationary, of zero mean, and are uniformly bounded (this is merely for simplicity of analysis, so that all the moments of these signals exist). Define $r_m = Eu_k u_{k+m}$ (autocorrelations of u_k) and $c_m = Eu_k d_{k+m}$ (correlations between u_k and d_k). Let

$$R_u = \begin{bmatrix} r_0 & r_1 & \cdots & r_{n-1} \\ r_1 & r_0 & \cdots & r_{n-2} \\ \vdots & \vdots & \cdots & \vdots \\ r_{n-1} & r_{n-2} & \cdots & r_0 \end{bmatrix}, \quad B = \begin{bmatrix} c_0 \\ c_1 \\ \vdots \\ c_{n-1} \end{bmatrix}. \tag{7.9}$$

(A7.2) This is the same as assumption (A7.1), except that u_k and d_k are not necessarily independent.

Theorem 7.2 *Under* (A7.2), *the following limits hold with probability one as $N \to \infty$:*

$$\frac{1}{N}\sum_{k=1}^{N} \varphi_k \varphi_k' \to R_u,$$
$$\left(\frac{1}{N}\sum_{k=1}^{N} \varphi_k \varphi_k'\right)^{-1} \to R_u^{-1},$$
$$\frac{1}{N}\sum_{k=1}^{N} \varphi_k d_k \to B,$$

where R_u and R_u^{-1} are positive definite, and

$$\widehat{\theta}_N \to \widehat{\theta} = \theta + R_u^{-1} B. \tag{7.10}$$

Proof. See [72]. □

As a result of Theorem 7.2, accuracy of channel estimation depends critically on correlations between u_k and d_k. Using the time-split ANC can remedy this problem.

We provide an analysis on convergence rates of the estimates. It will be shown that not only do correlations result in an estimation bias as evidenced by Eq. (7.10); they also reduce convergence rates significantly. Again in this aspect, it becomes favorable to choose an interval of smaller correlation for system identification. To understand this, we start with the case of uncorrelated u_k and d_k. In this case, the large deviation theorems show that exponential convergence rates are guaranteed.

Uncorrelated noise and source: We concentrate on a special case with the following assumptions: Suppose that $\{u_k\}$ and $\{d_k\}$ satisfy (A7.1). In addition, $\{u_k\}$ is an i.i.d. sequence. Consider the error term $r_N = (\sum_{k=1}^{N} \varphi_k d_k)/N$. A typical component of r_N is $z_N^e = \frac{1}{N}\sum_{k=1}^{N} u_{k+m} d_k$. Define, for $\eta > 0$, $m(\eta) = \inf_{z \geq 0} \mathrm{e}^{-z\eta} E\{\mathrm{e}^{z(u_1 d_1)}\}$. The large deviations principle (3.2) ensures that for every $\eta > 0$,

$$P\left\{\left|\frac{1}{N}\sum_{k=1}^{N} u_{k+m} d_k\right| \geq \eta\right\} \leq (m(\eta))^N. \tag{7.11}$$

For every $0 < \kappa < m(\eta)$ and for sufficiently large N,

$$P\left\{\left|\frac{1}{N}\sum_{k=1}^{N} u_{k+m} d_k\right| \geq \eta\right\} \geq (m(\eta) - \kappa)^N. \tag{7.12}$$

Indeed, under the hypothesis, u_k and d_k are independent. As a result, $u_{k+m} d_k$ have zero mean and are i.i.d. The inequalities (7.11) and (7.12) follow from Theorem 4.3.

Remark 7.3 Equation (7.11) shows that the estimation error due to noise converges to zero exponentially. Equation (7.12) indicates how tight the exponential bounds can be. The moment-generating function $E\{e^{z(u_1 d_1)}\}$ depends on the actual distributions of u_1 and d_1. If $u_1 d_1$ is normally distributed, then $m(\eta) = \exp(-\eta^2/(2\sigma^2))$. It follows that in this case,

$$\begin{aligned} &\kappa_1 \exp\left(-\frac{\eta^2}{2\sigma^2} - \kappa\right)^N \\ &\leq P\left\{\left|\frac{1}{N}\sum_{k=1}^{N} u_{k+m} d_k\right| \geq \eta\right\} \\ &\leq \kappa_2 \exp\left(-\frac{\eta^2 N}{2\sigma^2}\right). \end{aligned} \tag{7.13}$$

Furthermore, by the central limit theorem, r_N is asymptotically normally distributed. As a result, for sufficiently large N, the convergence rates in Eq. (7.13) provide a good approximation even when the underlying distributions are not normal. These results show that if signal and noise are independent, the ANC is a very good identification scheme, since the probability of large estimation errors is exponentially small.

Correlated noises and source: When u_k and d_k are correlated, a typical term

$$\frac{1}{N}\sum_{k=1}^{N} u_{k+m} d_k$$

converges to c_m w.p.1. Define $w_k = u_{k+m} d_k - c_m$. Then w_k satisfies $Ew_k=0$. Now for $l \neq 0$,

$$\gamma_l = Ew_k w_{k+l} = E(u_{k+m} d_k - c_m)(u_{k+m+l} d_{k+l} - c_m)$$

may not be zero. Hence in general, w_k is not uncorrelated. However, if the correlation is of short memory, namely γ_l is small in a certain sense, then convergence rates can be established.

(A7.3) (1) Assumption (A7.2) is valid. (2) γ_l satisfies

$$b = \lim_{n\to\infty} \sum_{l=-n}^{n} \gamma_l (1 - |l|/n) < \infty.$$

Define $S_N = \sum_{k=1}^{N} w_k/N$. Assume the existence of the following moment-generating functions and their limits:

$$\Gamma_N(\lambda) =: \log E e^{\lambda S_N}, \quad \Gamma(\lambda) = \lim_{N\to\infty} \frac{1}{N}\Gamma_N(N\lambda). \tag{7.14}$$

Let $\Gamma^*(x)$ be the Fenchel–Legendre transform of $\Gamma(\lambda)$,

$$\Gamma^*(x) = \sup_{\lambda\in\mathbb{R}} [\lambda x - \Gamma(\lambda)].$$

Theorem 7.4 *Under* (A7.3) *and the existence of the function in* (7.14),

$$\limsup_{N\to\infty} \frac{1}{N} \log P\{\|S_N\| \ge \eta\} \le -\frac{\eta^2}{2b}.$$

Proof. See Theorem 5.1. □

Theorems 7.2 and 7.4 provide a foundation for analyzing the benefits of our method. Consider the source signals in emergence and nonexistence stages cyclicly, which will be denoted by d^e and d^n, respectively. Correspondingly, measured source signals are $x^e = d^e + Gu$, $x^n = d^n + Gu$. Our method utilizes x^n for identification and x^e for noise cancellation. In both cases, the inputs are the measured virtual noise u.

Under (A7.2), denote the correlation matrices in Eq. (7.9) by B^e for correlation between u and x^e, and B^n for correlation between u and x^n. The autocorrelation matrix R_u varies slightly between the two phases, and these will be denoted by R_u^e and R_u^n, respectively. The channel estimate from the ANC is $\widehat{\theta}_N^e = \theta + (R_u^e)^{-1}B^e = \theta + r_N^e$, and the estimate from our method is $\widehat{\theta}_N^n = \theta + (R_u^n)^{-1}B^n = \theta + r_N^n$. Applying these estimates to noise cancellation on the process x^e, we have

$$\widehat{d}_k^{\rm anc} = x_k^e - \varphi_k'\widehat{\theta}_N^e = d_k^e + \varphi_k'\theta - \varphi_k'\widehat{\theta}_N^e = d_k^e - \varphi_k' r_N^e$$

for the ANC, and $\widehat{d}_k^{\rm tsanc} = x_k^e - \varphi_k'\widehat{\theta}_N^n = d_k^e + \varphi_k'\theta - \varphi_k'\widehat{\theta}_N^n = d_k^e - \varphi_k' r_N^n$ for our method. Consequently, the errors in signal extractions are $\mathrm{e}_k^{\rm anc} = d_k^e - \widehat{d}_k^{\rm anc} = \varphi_k' r_N^e$ for the ANC, and

$$\mathrm{e}_k^{\rm tsanc} = d_k^e - \widehat{d}_k^{\rm tsanc} = \varphi_k' r_N^n$$

for our method. Define

$$\begin{aligned}
\|B^e\|_2 &= \alpha^e, \\
\|B^n\|_2 &= \alpha^n, \\
\overline{\sigma}(R_u^e)^{-1} &= \beta^e, \\
\overline{\sigma}(R_u^n)^{-1} &= \beta^n,
\end{aligned}$$

where $\overline{\sigma}$ is the largest singular value of a matrix.

Theorem 7.5 (1) *The sample means of* $(\mathrm{e}_k^{\rm anc})^2$ *and* $(\mathrm{e}_k^{\rm tsanc})^2$ *converge w.p.1 to* $(B^e)'(R_u^e)^{-1}B^e$ *and* $(B^n)'(R_u^n)^{-1}B^n$, *respectively.*

(2) *The limits are bounded by* $(B^e)'(R_u^e)^{-1}B^e \le \beta^e(\alpha^e)^2$, $(B^n)'(R_u^n)^{-1} B^n \le \beta^n(\alpha^n)^2$.

Proof. See [72]. □

Theorem 7.5 states that our method can reduce signal reconstruction errors, in terms of mean square errors, by a factor of at least

$$\eta = \beta^e(\alpha^e)^2/\beta^n(\alpha^n)^2.$$

In the special case of $\beta^n = \beta^e$, the factor becomes $\eta = (\alpha^e/\alpha^n)^2$.

8

Applications to Electric Machines

Electric machines are essential systems in electric vehicles and are widely used in other applications. In particular, permanent magnet direct current (PMDC) motors have been extensively employed in industrial applications such as electric vehicles and battery-powered devices such as wheelchairs, power tools, guided vehicles, welding equipment, X-ray and tomographic systems, and computer numerical control (CNC) machines. PMDC motors are physically smaller and lighter for a given power rating than induction motors. The unique features of PMDC motors, including their high torque production at lower speed and flexibility in design, make them preferred choices in automotive transmissions, gear systems, lower-power traction utility, and other fields [12, 18, 22, 47, 54]. For efficient torque/speed control, thermal management, motor-condition monitoring, and fault diagnosis of PMDC motors, it is essential that their characteristics be captured in real-time operations. This is a system identification problem that can be carried out by using standard identification methods [35, 48].

This section is a case study in identification of PMDC motor parameters using only binary-valued sensors. In general, binary sensors are preferred in practical systems because they are more cost-effective than regular sensors. In many applications they are the only ones available for real-time operations. For remotely controlled motors, binary sensing can reduce communication costs dramatically. Using binary sensors is challenging for system modeling, identification, and control, since they are nonlinear, discontinuous, and provide very limited information. The main methodology follows

Q. He et al., *System Identification Using Regular and Quantized Observations*, SpringerBriefs in Mathematics, DOI 10.1007/978-1-4614-6292-7_8,

schemes based on stochastic empirical measure identification developed in [60, 62, 63].

8.1 Identification of PMDC-Motor Models

Typical models for a DC motor contain one differential equation for the electric part, one differential equation for the mechanical part, and their interconnections. This is valid for PMDC motors as well; see Fig. 8.1.

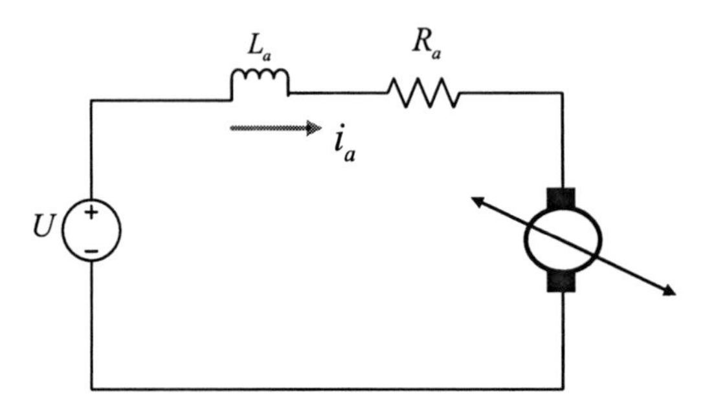

FIGURE 8.1. A PMDC motor

The equation for the motor rotor and shaft motions is

$$\frac{\mathrm{d}w(t)}{\mathrm{d}t} = \frac{-F}{J}\omega(t) + \frac{k}{J}i_a(t) - \frac{1}{J}T_L(t), \tag{8.1}$$

and for the stator wiring is

$$\frac{\mathrm{d}i_a(t)}{\mathrm{d}t} = \frac{-k}{L_\mathrm{a}}\omega(t) - \frac{R_\mathrm{a}}{L_\mathrm{a}}i_a(t) + \frac{1}{L_\mathrm{a}}U(t), \tag{8.2}$$

where $\omega(t)$ is the shaft speed (rad/sec), $i_a(t)$ the motor current (A), $U(t)$ the supply voltage (V), $T_\mathrm{L}(t)$ the load torque (N·m), J the moment of inertia for the motor (Kg·m^2), F the friction coefficient (N·m·s), k the motor constant (N·m/A or V/rad/s), L_a the armature inductance (henrys), and R_a the armature resistance (ohms).

The differential equations (8.1) and (8.2) can be transformed into a minimum state space realization (both controllable and observable) as

$$\begin{aligned} \dot{x}(t) &= A_0 x(t) + B_0 u(t), \\ y(t) &= C_0 x(t), \end{aligned}$$

where $x(t) = [\omega(t), i_a(t)]'$ and $u(t) = [T_{\rm L}(t), U(t)]'$ are the state and input vectors, respectively; $y(t)$ is the output; and

$$A_0 = \begin{bmatrix} \frac{-F}{J} & \frac{k}{J} \\ \frac{-k}{L_{\rm a}} & \frac{-R_{\rm a}}{L_{\rm a}} \end{bmatrix}, \quad B_0 = \begin{bmatrix} \frac{-1}{J} & 0 \\ 0 & \frac{1}{L_{\rm a}} \end{bmatrix}, \quad C_0 = \begin{bmatrix} 1 & 0 \\ 0 & 1 \end{bmatrix}.$$

The transfer matrix of the system $G(s)$ from the input u to the output y is

$$G(s) = C_0(sI - A_0)^{-1}B_0.$$

The transfer function between the input voltage and the angular speed is

$$\frac{\Omega(s)}{U(s)} = \frac{k}{(Js + F)(L_{\rm a}s + R_{\rm a}) + k^2}$$

$$= \frac{k/L_{\rm a}J}{s^2 + \left(\frac{R_{\rm a}}{L_{\rm a}} + \frac{F}{J}\right) + \frac{R_{\rm a}F + k^2}{L_{\rm a}J}}.$$

Under a given sampling interval with a zero-order hold function, the corresponding discrete-time transfer function in z-transform can be generically expressed as a second-order system

$$\omega_k = G(q)u_k = \frac{b_1 q + b_2 q^{-2}}{1 + a_1 q + a_2 q^{-2}} u_k,$$

where q is the one-step shift operator $qu_k = u_{k-1}$.

For system identification experiments, it is desirable to transform a model into a regression form. For the speed system, we have the general autoregressive moving average (ARMA) model [35, 48]

$$\omega_k = \sum_{j=1}^{n} -a_j \omega_{k-j} + \sum_{i=0}^{m} b_i u_{k-i}.$$

In our case, the identified system of the PMDC motor has the following second-order ARMA model:

$$\omega_k = \varphi_k' \theta, \tag{8.3}$$

where ω_k is the speed (rad/sec), u_k the input voltage (V),

$$\varphi_k' = [-\omega_{k-1}, -\omega_{k-2}, u_{k-1}, u_{k-2}],$$

$$\theta' = [a_1, a_2, b_1, b_2],$$

and where θ is to be identified.

The output of the system is measured by a binary sensor with threshold C. The output of the sensor will be either 0 or 1, according to the following relation by an indicator function:

$$s_k = \chi_{\{y_k \leq C\}} = \begin{cases} 1, y_k \leq C, \\ 0, y_k \geq C. \end{cases}$$

The sensor may be a physical sensor such as a Hall effect sensor or a coding scheme if the output must be transmitted through a communication network.

(A8.1) Suppose that $\{d_k\}$ is a sequence of i.i.d. (independent and identically distributed) random variables. The accumulative distribution function $F(\cdot)$ of d_1 is twice continuously differentiable. The moment-generating function of d_1 exists. The inverse of the function $F(\cdot)$ exists and is $F^{-1}(\cdot)$.

The identification algorithm we will develop utilizes periodic input dithers, which are designed to simplify identification problems, provide persistent excitation, and make it possible to use the law of large numbers directly in achieving parameter convergence. Select u_k to be four-periodic and of full rank; see [62] for detailed definitions and discussion. We take N samples of the sensor output. For convenience of notation, assume that N is a multiple of the size of θ, which is 4. Hence, let the observation length be $N = 4L$ for some positive integer L. Then the noise-free system output $\omega_k = G(q)u_k$ is also four-periodic after a short transient duration, since the system is exponentially stable.

8.2 Binary System Identification of PMDC Motor Parameters

A unique idea of binary system identification is that stochastic information from measurement noise can be used to our advantage in overcoming the fundamental limitations of the binary sensor. The sensor output does not give any accurate information about the system output. Interestingly, by using the noise features and the law of large numbers, more accurate information of the system output can be asymptotically obtained from the 0, 1 sequence of the sensor output.

The system output is either corrupted by the measurement noise or expanded with an added dither of zero mean and standard deviation σ. The output becomes $y_k = \omega_k + d_k = G(q)u_k + d_k$. For each $j = 1, \ldots, 4$, we first find the estimated γ_L^j of some intermediate variables γ^j by

$$\gamma_L^j = C - F^{-1}(\zeta_L^j),$$

where $\zeta_L^j = \frac{1}{L}\sum_{l=0}^{L-1} s_{j+2ln}$ is the sample average of the sensor output.

Define an estimated four-periodic output sequence of $G(q)$ by the estimated γ^j as $\widehat{\omega}_{j+4l} = \widehat{\omega}_j = \widehat{\gamma}^j$, $j = 1, \ldots, 4$ and $l = 1, \ldots, L-1$. To estimate the parameter θ, we use $\widehat{\omega}_k$ in place of ω_k, $\widehat{\omega}_k = \widehat{\varphi}'_k \widehat{\theta}_L$, where $\widehat{\varphi}'_k = [\widehat{\omega}_{k-1}, \widehat{\omega}_{k-2}, u_{k-1}, u_{k-2}]$. By defining $\widehat{Y} = [\widehat{\omega}_1, \ldots, \widehat{\omega}_4]'$, $\widehat{\Phi} = [\widehat{\varphi}_1, \widehat{\varphi}_2, \widehat{\varphi}_3, \widehat{\varphi}_4]'$, we have

$$\widehat{\Omega} = \widehat{\Phi}\widehat{\theta}_L.$$

When the regression matrix Φ is of full rank, one derives an estimate θ_L from

$$\widehat{\theta}_L = \widehat{\Phi}^{-1}\widehat{\Omega},$$

or

$$\widehat{\theta}_L = (\widehat{\Phi}'\widehat{\Phi})^{-1}\widehat{\Phi}'\widehat{\Omega}.$$

8.3 Convergence Analysis

The convergence of the above algorithm was established in [60, 62].

Theorem 8.1 ([60, Theorem 3]) *Suppose that $G(q) = \frac{D(q)}{B(q)}$, and $D(q)$ and $B(q)$ are coprime polynomials, i.e., they do not have common roots. If $\{u_k\}$ is $2n$-periodic and of full rank, then $\widehat{\theta}_L \to \theta$ w.p.1 as $L \to \infty$.*

Example 8.2 Suppose that the PMDC motor has the parameters $L_a = 0.0104\,\mathrm{H}$, $R_a = 1.43\,\mathrm{ohm}$, $J = 0.068\,\mathrm{Kg \cdot m^2}$, $F = 0.0415\,\mathrm{N \cdot m \cdot s}$, $k = 1.8\,\mathrm{N \cdot m/A}$ or V/rad/s. The system is simulated over the time interval $[0, 2]$ seconds with sampling period $T = 0.01\,\mathrm{s}$. The nominal input voltage is 240 V. The regression model has the true parameters $a_1 = -1.0078$, $a_2 = 0.2513$, $b_1 = 0.0815$, $b_2 = 0.0514$:

$$\omega_k = -a_1\omega_{k-1} - a_2\omega_{k-2} + b_1 u_{k-1} + b_2 u_{k-2}.$$

In this experiment, we apply a 4-periodic input voltage to a PMDC motor model with its first four values $[240, 216, 264, 204]$ V.

The binary sensor threshold is $C = 125$.[1] The measurement noises are independent and identically distributed Gaussian noise sequences with zero mean and standard deviation $\sigma = 4$. Figure 8.2 shows the estimated and actual motor speeds.

---

[1] In fact, in the actual implementation of the algorithm, the threshold is adaptively selected to improve estimation accuracy. This value is selected as an example that is close to the average of the periodic output values.

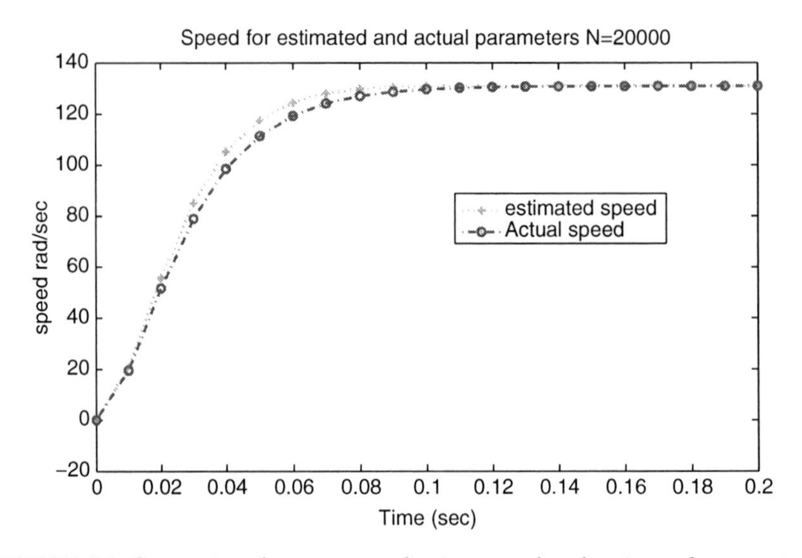

FIGURE 8.2. Comparison between speed using actual and estimated parameters

8.4 Quantized Identification

When more complicated quantized sensors are used, more information can be extracted from the sensor output, which can potentially improve estimation accuracy. However, proper use of the sensor information is far from a trivial issue. We will use a scheme that is optimal in the sense that the estimation error variance is minimized.

Suppose that now we have a quantized sensor with m thresholds $-\infty < C_1 < \cdots < C_m < \infty$, and the sensor output can be represented by a set of m indicator functions $s_k = [s_k(1), \ldots, s_k(m)]'$, where $s_k(i) = \chi_{\{-\infty < y_k < C_i\}}$, $i = \{1, \ldots, m\}$. First, we observe that for each threshold C_i, it is a binary sensor and $s_k(i)$ is the corresponding sensor output. Consequently, all discussions in the previous section on binary sensors are valid, including input design, algorithms, convergence properties, and impact of threshold selections. Since these binary sensors provide information on the same ω_j, $j = 1, 2, 3, 4$, the main issue here is how to combine information from these binary sensors of different thresholds to form a new combined estimate of the same quantity.

It is obvious that each threshold C_i can generate an estimate of ω. A suitable combination of these estimates will lead to an asymptotically optimal estimator for θ by achieving the Cramér–Rao lower bound.

Define the weighting $\gamma = [\gamma_1, \ldots, \gamma_m]$ such that $\gamma_1 + \cdots + \gamma_m = 1$. From the m estimates ω_N^i of ω using the m sensor thresholds, their convex combination is also an estimate $\widehat{\omega}$ of ω:

$$\widehat{\omega} = \sum_{i=1}^{m} \gamma_i \omega_N^i = \gamma' W_N,$$

where $W_N = [\omega_N^1, \dots, \omega_N^m]$. The value $\widehat{\omega}$ is called a quasiconvex combination estimator (QCCE). When the weighting values are selected optimally, we have the optimal QCCE; see [58].

The optimization algorithm is described below. Let ω_N^i, $i = 1, \dots, m$, be m asymptotically unbiased estimators of ω based on samples of size N. Then the estimation error is defined by $e_N^i = \omega_N^i - \omega$ for each $i = 1, \dots, m$. The error vector can be expressed as $e_N = \omega_N - \omega\mathbf{1}$, where $\mathbf{1} = [1, 1, \dots, 1]'$. Define the covariance matrix of e_N as $V_N = E[e_N e_N']$.

Theorem 8.3 *Suppose that $V_N(\omega)$ is positive definite. Then the optimal QCCE is obtained by choosing*

$$\gamma^* = \frac{V_N^{-1}(\omega)\mathbf{1}}{\mathbf{1}'V_N^{-1}(\omega)\mathbf{1}}.$$

The minimal variance is

$$\sigma_N^2 = \frac{1}{\mathbf{1}'V_N^{-1}(\omega)\mathbf{1}}.$$

One way to implement the QCCE numerically is as follows:

- Step 1. Find the sample mean of all estimated values $\widehat{\omega}$, computed from the m-thresholds. The sample mean is

$$\bar{W}_N = \sum_{j=1}^{N} W_j/N.$$

- Step 2. Find the sample covariance $\widehat{V}_N$. The sample covariance is

$$\widehat{V}_N = \frac{1}{N-1}\sum_{j=1}^{N}(W_j - \bar{W}_N)(W_j - \bar{W}_N)'.$$

- Step 3. Find γ_N as

$$\gamma_N = \frac{\widehat{V}_N^{-1}\mathbf{1}}{\mathbf{1}'\widehat{V}_N^{-1}\mathbf{1}}.$$

- Step 4. Find $\widehat{\omega}_N = (\gamma_N)'W_N$.

Example 8.4 We consider the same system as in Example 8.2, with a fixed noise standard deviation $\sigma = 4$, and the observation length $N = 5000$, but with different threshold values. The results are shown in Table 8.1. Note that the missing entry in Table 8.1 is due to data limitations. In the table, TSE is an abbreviation for Total Squared Error.

TABLE 8.1. Parameter estimation using individual thresholds with $M = 240$ V, $N = 5000$, and $\sigma = 4$

Parameter	Actual value	Estimated C=115	Estimated C=122	Estimated C=130	Estimated C=135
a_1	−1.0078	−0.8509	−0.8777	−0.8853	−0.9398
a_2	0.2513	0.0252	0.136	0.2241	0.2891
b_1	0.0815	0.0948	0.0901	0.1093	0.0909
b_2	0.0514	0.0481	0.0395	0.0551	0.0571
TSE		0.07593	0.030436	0.016533	0.006174

Theorem 8.5 ([58]) *The optimal QCCE is asymptotically efficient in the sense that*

$$N\sigma_N^2 - N\sigma_{CR}^2(N,m) \to 0 \quad as \quad N \to \infty.$$

Example 8.6 This example involves a quantized sensor. We consider the same system as in Example 8.2. The quantized sensor has four thresholds $C_1 = 110, C_2 = 118, C_3 = 123, C_4 = 133$. The measurement noises are i.i.d. Gaussian noise sequences with zero mean and standard deviation $\sigma = 30$. The QCCE is used for speed estimation. Figure 8.3 shows the sample variance and the theoretical CR bound.

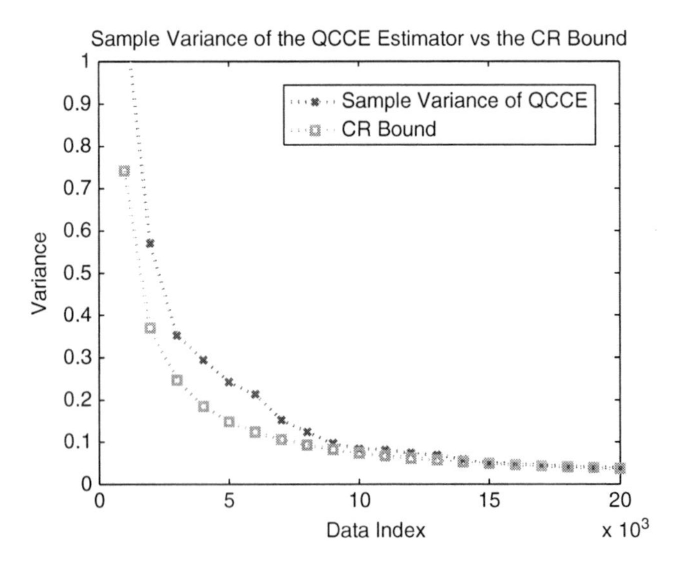

FIGURE 8.3. Sample variance of the QCCE estimator vs. the CR lower bound

8.5 Large Deviations Characterization of Speed Estimation

The following example demonstrates the large deviations characterization of the estimation errors.

Example 8.7 The first case involves a binary-valued sensor. Theorems 4.6 and 4.7 are applicable in this case. Consider the same system as in Example 8.2. The system equation is

$$\omega_k = -a_1\omega_{k-1} - a_2\omega_{k-2} + b_1 u_{k-1} + b_2 u_{k-2},$$

and $a_1 = -1.0078$, $a_2 = 0.2513$, $b_1 = 0.0815$, $b_2 = 0.0514$. Under the four-periodic input voltage with the first period $[240, 216, 264, 204]$, we can calculate that the true values of ω_k after reaching its periodicity are

$$\gamma_1 = 126.4180, \quad \gamma_2 = 125.6808, \quad \gamma_3 = 125.0175, \quad \gamma_4 = 127.1942.$$

The sensor threshold is $C = 125$. Since this model does not involve unmodeled dynamics, $\widetilde{C}_i = \gamma_i$, $i = 1, 2, 3, 4$. As a result, $\widetilde{c}_1 = C - \gamma_1 = -1.4180$, $\widetilde{c}_2 = C - \gamma_2 = -0.6808$, $\widetilde{c}_3 = C - \gamma_3 = -0.0175$, $\widetilde{c}_4 = C - \gamma_4 = -2.1942$.

Under the i.i.d. Gaussian noise sequences with zero mean and standard deviation $\sigma = 4$, $b_1 = P\{d(t_0) \le \widetilde{c}_1\} = 0.3615$, $b_2 = P\{d(t_0) \le \widetilde{c}_2\} = 0.4324$, $b_3 = P\{d(t_0) \le \widetilde{c}_4\} = 0.4983$, $b_4 = P\{d(t_0) \le \widetilde{c}_4\} = 0.2917$. Consequently, the rate function is

$$\begin{aligned} I(\beta) &= \sum_{i=1}^{4} \log \frac{\beta_i^{\beta_i}(1-b_i)^{\beta_i-1}}{b_i^{\beta_i}(1-\beta_i)^{\beta_i-1}} \\ &= \log \frac{\beta_1^{\beta_1}(1-0.3615)^{\beta_1-1}}{0.3615^{\beta_1}(1-\beta_1)^{\beta_1-1}} + \log \frac{\beta_2^{\beta_2}(1-0.4324)^{\beta_2-1}}{0.4324^{\beta_2}(1-\beta_2)^{\beta_2-1}} \\ &\quad + \log \frac{\beta_3^{\beta_3}(1-0.4983)^{\beta_3-1}}{0.4983^{\beta_3}(1-\beta_3)^{\beta_3-1}} + \log \frac{\beta_4^{\beta_4}(1-0.2917)^{\beta_4-1}}{0.2917^{\beta_4}(1-\beta_4)^{\beta_4-1}}, \end{aligned}$$

where $0 \le \beta_i \le 1$, $i = 1, \ldots, 4$.

TABLE 8.2. Actual and estimated parameters

Parameters	Actual value	Estimated N =1,000	Estimated N =5,000	Estimated N =10,000	Estimated N =20,000
a_1	−1.0078	−0.8628	−0.9004	−1.0246	−1.0079
a_2	0.2513	0.2848	0.3123	0.2866	0.2723
b_1	0.0815	0.09378	0.07201	0.0882	0.0872
b_2	0.0514	0.0346	0.0421	0.0547	0.0571
Error norm		0.0225803	0.01543	0.001584	0.000506

Numerical calculations of estimation errors can be performed from the above rate function using the large deviations principle. Simulation results on speed estimation errors as functions of the sample sizes are shown in Table 8.2. Note that the missing entry in Table 8.2 is due to data limitations.

9

Remarks and Conclusion

9.1 Discussion of Aperiodic Inputs

For clarity, this book assumes that the input is designed to be periodic. Advantages of using full rank and periodic inputs in quantized identification problems have been extensively discussed in [62]. Note that such a choice is not a fundamental limitation, and aperiodic inputs can also be used. As a first attempt at using LDPs in complexity analysis, we will not explore this direction in detail here, but only highlight the main steps. To illustrate the basic ideas and computational processes for quantized identification under aperiodic inputs, we use the basic case of a binary sensor with threshold C and gain identification problems. In this case, the system is $y(t) = \theta u(t) + d(t)$, for $t = 0, 1, \ldots$, and $s(t) = I_{\{y(t) \leq C\}}$, where θ is to be identified. The empirical measure on the basis of k measurements is

$$\xi_k = \frac{1}{k} \sum_{t=0}^{k-1} s(t).$$

Set

$$\begin{aligned} G_k(\theta) &= E\xi_k = \frac{1}{k} \sum_{t=0}^{k-1} Es(t), \\ Es(t) &= \frac{1}{k} \sum_{t=0}^{k-1} F(C - \theta u(t)). \end{aligned}$$

Q. He et al., *System Identification Using Regular and Quantized Observations*, SpringerBriefs in Mathematics, DOI 10.1007/978-1-4614-6292-7_9,

Then, the estimate θ_k of θ is $\theta_k = G_k^{-1}(\xi_k)$. Although explicit expressions for $G(\cdot)$ and $G^{-1}(\cdot)$ may be difficult to obtain, its numerical solutions are straightforward. Note that

$$\begin{aligned} \varepsilon_k &= \xi_k - E\xi_k = \frac{1}{k}\sum_{t=0}^{k-1}(s(t) - F(C - \theta u(t))) \quad \text{and} \\ e_k &= \theta_k - \theta = G^{-1}(\xi_k) - G^{-1}(E\xi_k), \end{aligned}$$

which may be approximated for small ε_k by

$$e_k \approx \left(\frac{\partial G(\theta)}{\partial \theta}\right)^{-1} \varepsilon_k := g_k(\theta)\varepsilon_k,$$

where

$$\begin{aligned} \frac{\partial G_k(\theta)}{\partial \theta} &= \frac{1}{k}\sum_{t=0}^{k-1} \frac{\partial F(C - \theta u(t))}{\partial \theta} \\ &= \frac{1}{k}\sum_{t=0}^{k-1} \widetilde{v}(y), \\ \widetilde{v}(t) &= -u(t) f(C - \theta u(t)). \end{aligned}$$

Since the LDP is preserved by a continuous mapping, we need only concentrate on ε_k (whose rate function can be easily modified by $g_k(\theta)$ to obtain the rate function of e_k). However, ε_k is a convex combination of independent variables. Hence, the rate function of ε_k can be obtained from those of $s(t) - F(C - \theta u(t))$. When $u(t)$ is periodic (in this special case, it is a constant), these variables become also identically distributed, affording a substantial simplification of the computation. This, however, does not alter the fundamental properties of LDPs and rate function expressions. In the special case of uniform distributions, more explicit expressions and much easier computations can be obtained. Indeed, if $d(t)$ is i.i.d., uniformly distributed in $[-\delta, \delta]$, we have $F(x) = \frac{1}{2\delta}(x + \delta)$ and $f(x) = \frac{1}{2\delta}, x \in [-\delta, \delta]$. Consequently, assuming $C - \theta u(t) \in [-\delta, \delta]$ for all t, we have with $\overline{u}_k = \sum_{t=0}^{k-1} u(t)/k$,

$$\begin{aligned} G_k(\theta) &= \frac{1}{k}\sum_{t=0}^{k-1} \frac{1}{2\delta}(C - \theta u(t) + \delta) \\ &= \frac{C + \delta}{2\delta} - \frac{\frac{1}{k}\sum_{t=0}^{k-1} u(t)\overline{u}_k}{2\delta}\theta, \\ \theta_k &= \frac{\frac{C+\delta}{2\delta} - \xi_k}{\frac{\overline{u}_k \frac{1}{k}\sum_{t=0}^{k-1} u(t)}{2\delta}}, \quad \frac{\partial G_k(\theta)}{\partial \theta} \\ &= -\frac{1}{2\delta}\frac{1}{k}\sum_{t=0}^{k-1} u(t)\overline{u}_k, \\ e_k &\approx -\frac{2\delta}{\overline{u}_k \frac{1}{k}\sum_{t=0}^{k-1} u(t)}\varepsilon_k. \end{aligned}$$

Consequently, the rate function computation becomes straightforward.

9.2 Escape from a Domain

In this section, we provide an alternative viewpoint for the study of large deviations in parameter estimates. Concentrating on the identification problem with binary observations, in contrast to the study in the previous sections, rather than dealing with the discrete processes directly, we take a continuous-time interpolation. Consider the algorithm

$$Z_k^i = \frac{1}{k}\sum_{l=0}^{k-1} \chi_{\{y(lm_0+i)\le C\}}. \tag{9.1}$$

For ease of discussion, assume that there is no unmodeled dynamics. Thus, $Y_l = \Phi_0\theta + D_l$. Define $\chi_k^i = \chi_{\{d(km_0+i)\le C-(\Phi_0\theta)_i\}}$ and

$$\begin{aligned} S_k &= (\chi_k^1,\ldots,\chi_k^{m_0})', \\ Z^* &= (F(C-(\Phi_0\theta)_1),\ldots,F(C-(\Phi_0\theta)_{m_0}))', \end{aligned} \tag{9.2}$$

where $(\Phi_0\theta)_i$ denotes the ith component of $\Phi_0\theta$. Define $Z_k = (Z_k^1,\ldots,Z_k^{m_0})'$. We can then write Eq. (9.1) recursively as

$$Z_{k+1} = Z_k - \frac{1}{k+1}Z_k + \frac{1}{k+1}S_k \begin{bmatrix} \chi_{\{d(km_0+1)\le C-(\Phi_0\theta)_1\}} \\ \vdots \\ \chi_{\{d(km_0+m_0)\le C-(\Phi_0\theta)_{m_0}\}} \end{bmatrix}. \tag{9.3}$$

Rather than directly analyzing the sequence of iterates, we use the method of ordinary differential equations for stochastic approximation; see Kushner and Yin [31, Chaps. 5 and 6]. Define $t_k = \sum_{l=0}^{k-1} 1/(l+1)$ and $m(t) = \max\{k : t_k \le t\}$, where t_k connect the discrete iteration number with continuous time, and $m(t)$ serves as its inverse. Define the piecewise constant interpolation

$$Z^{\{0\}}(t) = Z_k \quad \text{for} \quad t \in [t_k, t_{k+1}) \quad \text{and} \quad Z^{\{k\}}(t) = Z^{\{0\}}(t+t_k).$$

Suppose, for simplicity, that $\{d_k\}$ is an i.i.d. sequence with zero mean and finite variance. We can verify that w.p.1, $Z^{\{k\}}(\cdot)$ is uniformly bounded and equicontinuous in the extended sense (see [31, p. 102]). The Arzelà–Ascoli theorem [31, p. 102] yields that every convergent subsequence of $Z^{\{k\}}(\cdot)$ has a limit $Z(\cdot)$, which is a solution of the ordinary differential equation

$$\dot{Z}(t) = -Z(t) + Z^* \begin{bmatrix} F(C-(\Phi_0\theta)_1) \\ \vdots \\ F(C-(\Phi_0\theta)_{m_0}) \end{bmatrix}. \tag{9.4}$$

Moreover, as $k \to \infty$, $Z^{\{k\}}(\cdot + T_k) \to Z^*$ w.p.1, where $\{T_k\}$ is any sequence of positive real numbers satisfying $T_k \to \infty$.

We proceed to study the asymptotic properties of $Z^{\{k\}}(\cdot)$. Of particular interest are estimates of the probabilities of $Z^{\{k\}}(\cdot)$ escaping from a fixed neighborhood of the stable point Z^* of Eq. (9.4). To be precise, let G be a neighborhood of Z^* and define

$$\widetilde{\tau}_G^k = \min\{t : Z^{\{k\}}(t) \notin G\}.$$

By the w.p.1 convergence, the probability $P_x^k\{\widetilde{\tau}_G^k \leq T\}$ tends to zero as $k \to \infty$, and it is natural to look for the rate of convergence. In particular, we seek a sequence $\lambda_k \to 0$ and $0 < V_2 < V_1 < \infty$ such that the following limit exists:

$$\begin{aligned} -V_1 &\leq \liminf_{k\to\infty} \lambda_k \log P_x^k\{\widetilde{\tau}_G^k \leq T\} \\ &\leq \limsup_{k\to\infty} \lambda_k \log P_x^k\{\widetilde{\tau}_G^k \leq T\} \\ &\leq -V_2, \end{aligned}$$

where P_x^k denotes the probability conditioned on the event that $Z^{\{k\}}(0) = x \in G$. To this end, we need the following assumption.

Assume that there exist a function $H(\cdot,\cdot,\cdot) : \mathbb{R}^{m_0} \times \mathbb{R}^{m_0} \times [0,T] \to \mathbb{R}$ and a sequence $\lambda_k \to 0$ such that for every $x \in \mathbb{R}^{m_0}$ and piecewise constant function $\alpha(\cdot) : [0,T] \to \mathbb{R}^{m_0}$,

$$\int_0^T H(x, \alpha(s), s)ds = \lim_{k\to\infty} \lambda_k \log E \exp\left(\frac{1}{\lambda_k}\widetilde{\Pi}_k\right), \tag{9.5}$$

where

$$\widetilde{\Pi}_k = \sum_{i=0}^{N-1}\Big\langle \alpha'(i\Delta), \sum_{j=m(t_k+t)}^{m(t_k+i\Delta+\Delta-1)} \frac{1}{j+1}(-x+S_j)\Big\rangle$$

with S_j defined in Eq. (9.2) and $T = N\Delta$, and $\alpha(\cdot)$ is constant on the interval $[k\Delta, (k+1)\Delta)$. Let $C_x[0,T]$ denote the space of $\mathbb{R}^{m_0}$-valued continuous functions on $[0,T]$ with initial value x, with the super-norm topology. Denote the interior and closure of a set $A \subset C_x[0,T]$ by A° and $\overline{A}$, respectively. Then by [29, Theorem 1] we have the large deviations estimate.

Theorem 9.1 *Under the above assumption, for each set* $A \subset C_x[0,T]$,

$$\begin{aligned} -\inf_{\varphi\in A^\circ} S(T,\varphi) &\leq \liminf_{k\to\infty} \lambda_k \log P_x\{Z^{\{k\}}(\cdot) \in A\} \\ &\leq \limsup_{k\to\infty} \lambda_k \log P_x\{Z^{\{k\}}(\cdot) \in A\} \\ &\leq -\inf_{\varphi\in\overline{A}} S(T,\varphi), \end{aligned}$$

where

$$\begin{aligned} S(T,\varphi) &= \int_0^T L(\varphi(s),\dot{\varphi}(s),s)\mathrm{d}s, \\ L(x,\beta,s) &= \sup_{\alpha\in\mathbb{R}^{m_0}} [\langle \alpha,\beta\rangle - H(x,\alpha,s)]. \end{aligned}$$

To get the escape time estimate, set

$$A = \{\varphi(\cdot) : \varphi(0) = x, \varphi(t) \notin G \text{ for some } t \le T\}.$$

Then by Theorem 9.1,

$$\begin{aligned} -\inf_{\varphi\in A^\circ} S(T,\varphi) &\le \liminf_{k\to\infty} \lambda_k \log P_x\{\tilde{\tau}_G^k \le T\} \\ &\le \limsup_{k\to\infty} \lambda_k \log P_x\{\tilde{\tau}_G^k \le T\} \\ &\le -\inf_{\varphi\in\overline{A}} S(T,\varphi). \end{aligned}$$

To bring out the dynamical systems aspect of the problem, we further examine the escape probability of the iterates away from a neighborhood of the true parameter θ. To do so, we define

$$\widehat{Z}^{\{k\}}(t) = \Phi_0^{-1} C \begin{bmatrix} 1 \\ \vdots \\ 1 \end{bmatrix} - \Phi_0^{-1} F^{-1}(Z^{\{k\}}(t)),$$

where

$$F^{-1}(Z^{\{k\}}(t)) = \begin{bmatrix} F^{-1}(Z_1^{\{k\}}(t)) \\ \vdots \\ F^{-1}(Z_1^{\{k\}}(t)) \end{bmatrix},$$

where $F^{-1}(\cdot)$ is the inverse of $F(\cdot)$. It is easy to see that $\widehat{Z}^{\{k\}} \to \theta$ as $k \to \infty$. Given a neighborhood $\widehat{G}$ of θ, define

$$\widehat{\tau}_{\widehat{G}}^k = \min\{t : \widehat{Z}^{\{k\}}(t) \notin \widehat{G}\}.$$

Then

$$P_x\{\widehat{\tau}_{\widehat{G}}^k \le T\} \to 0.$$

Recall that

$$A = \{\varphi(\cdot) : \varphi(0) = x, \varphi(t) \notin G, \text{some } t \le T\},$$

and define

$$F(v) = \begin{bmatrix} F(v_1) \\ \vdots \\ F(v_{m_0}) \end{bmatrix}$$

for every for $v \in \mathbb{R}^{m_0}$. By applying the contraction principle, Proposition 3.5, we have the following estimates of rate of convergence for the escape from a domain problem:

$$\begin{aligned} -\inf_{\varphi \in B^\circ} S(T,\varphi) &\leq \liminf_{k\to\infty} \lambda_k \log P_x\{\widehat{\tau}^k_{\widehat{G}} \leq T\} \\ &\leq \limsup_{k\to\infty} \lambda_k \log P_x\{\widehat{\tau}^k_{\widehat{G}} \leq T\} \\ &\leq -\inf_{\varphi\in \overline{B}} S(T,\varphi), \end{aligned}$$

where

$$B = \{F(C - \Phi_0\varphi(\cdot)), \varphi(\cdot) \in A\}.$$

In the previous chapters, we have taken a direct approach for obtaining the large deviations bounds. This section provides an alternative for the study of large deviations. One may can also consult [11, 41, 45] for further reading related to efficiency issues for recursive estimates.

9.3 Randomly Varying Parameters

In this book, we have treated system identification under both regular and quantized sensors. We have concentrated on systems in which the underlying parameters are fixed constants. It appears that time-varying parameters, in particular randomly varying parameters, can also be treated. Recently, randomly varying parameters and regime-switching models have drawn renewed attention with a wide range of applications; see [67, 70, 74, 75] among others. Especially in [66, 68], we treated identification and tracking problems involving a Markovian parameter under both regular and binary sensors.

Using the setup similar to that in Chap. 2, consider

$$y(t) = \varphi'(t)\theta(t) + \widetilde{\varphi}'(t)\widetilde{\theta} + d(t), \quad t = t_0, \dots, t_0 + N - 1, \tag{9.6}$$

where

$$\begin{aligned} \varphi(t) &= (u(t), u(t-1), \dots, u(t-m_0+1))', \\ \widetilde{\varphi}(t) &= (u(t-m_0), u(t-m_0-1), \dots)'. \end{aligned}$$

The main difference between the model above and that in Eq. (2.3) is that the parameter θ is no longer a constant, but a random process. We may assume that the time-varying process $\{\theta(t)\}$ is a discrete-time Markov chain with a transition probability matrix P and a finite state space

$$\mathcal{M} = \{\theta^{(1)}, \theta^{(2)}, \dots, \theta^{(m_0)}\}.$$

In [66], we considered identification of the aforementioned systems with observation under regular sensors, whereas in [68], we considered binary-valued observations; see also background materials in [69]. Using the techniques developed in this book, we can proceed to develop the associated large deviations principles. Careful analysis is needed for carrying out the investigation.

9.4 Further Remarks and Conclusions

This book is devoted to the large deviations principle of system identification under regular and quantized observations. The results provide some unique characterizations of identification accuracy in probability in relation to quantization levels and data window sizes, and are of importance for a wide spectrum of applications that require statistical reliability assessment of diagnosis, decisions, control, and information processing. The LDP of these identification estimates confirms that the orders of convergence in probability are of exponential types and yields the precise rate functions. By establishing monotonicity of the rate functions with respect to the space complexity, the complexity results of this book can serve as a benchmark criterion for comparing space complexity of system identifications under binary, quantized, and regular sensors. Although the regular sensor shows more accuracy, it carries a much higher complexity in data flow rates, and hence requires more system resources. Consequently, a tradeoff can be made on the basis of such a complexity analysis. To highlight the possibilities of a wide range of applications, three chapters have been devoted to case studies of applications. We hope that the results presented and the examples given will motivate further investigation for a wider range of applications.

References

[1] K. Aström and B. Wittenmark, *Adaptive Control*, Addison-Wesley, 1989.

[2] R.R. Bahadur, Large deviation of the maximum likelihood estimate in the Markov chain case, in *Recent Advances in Statistics*, M.H. Rizvi, J.S. Rostag, and D. Siegmund eds., Academic Press, NY, 1983, 273–286.

[3] R.R. Bahadur, S.L. Zabell, and J.C. Gupta, Large deviations, tests and estimates, in *Asymptotic Theory of Statistical Tests*, I.M. Chakravarti, ed., Academic Press, NY, 1980, 33–64.

[4] C. Barbier, H. Meyer, B. Nogarede, S. Bensaoud, A battery state of charge indicator for electric vehicle, in *Proc. International Conference of the Institution of Mechanical Engineers, Automotive Electronics*, London, UK, May 17–19, 29–34, 1994.

[5] E. Barsoukov, J. Kim, C. Yoon, H. Lee, Universal battery parameterization to yield a non-linear equivalent circuit valid for battery simulation at arbitrary load, *J. Power Sour.*, 83 (1999), 61–70.

[6] B.S. Bhangu, P. Bentley, D.A. Stone, and C.M. Bingham, Nonlinear observers for predicting state-of-charge and state-of-health of lead-acid batteries for hybrid-electric vehicles, *IEEE Trans. Veh. Tech.*, 54 (2005), 783–794.

[7] W. Bryc and A. Dembo. Large deviations and strong mixing, *Ann. Inst. H. Poincare Probab. Stat.*, 32 (1996), 549–569.

[8] P.E. Caines, *Linear Stochastic Systems*, Wiley, New York, 1988.

Q. He et al., *System Identification Using Regular and Quantized Observations*, SpringerBriefs in Mathematics, DOI 10.1007/978-1-4614-6292-7,

[9] H. Chan, D. Sutanto, A new battery model for use with battery energy storage systems and electric vehicle power systems, in *Proceedings of the 2000 IEEE Power Engineering Society Winter Meeting*, Singapore, 470–475, January 23–27, 2000.

[10] H.-F. Chen and L. Guo, *Identification and Stochastic Adaptive Control*, Birkhäuser, Boston, 1991.

[11] K.L. Chung, On a stochastic approximation method, *Ann. Math. Statist.*, 25 (1954), 463–483.

[12] O. Craiu, A. Machedon, et al., 3D finite element thermal analysis of a small power PM DC motor, *12th International Conference on Optimization of Electrical and Electronic Equipment (OPTIM)*, 2010.

[13] H.F. Dai, X.Z. Wei, Z.C. Sun, Online SOC estimation of high-power lithium-ion batteries used on HEVs, *Proceedings of IEEE ICVES* 2006, 342–347.

[14] A. Dembo and O. Zeitouni, *Large Deviations Techniques and Applications*, 2nd ed., Springer-Verlag, New York, 1998.

[15] P. Dupuis and H.J. Kushner, Stochastic approximation via large deviations: asymptotic properties, *SIAM J. Control Optim.*, 23 (1985) 675–696.

[16] P. Dupuis and H.J. Kushner, Stochastic approximation and large deviations: upper bounds and w.p.1 convergence, *SIAM J. Control Optim.*, 27 (1989) 1108–1135.

[17] W. Feller, *An Introduction to Probability Theory and Its Applications*, vol. I, 3rd ed. Wiley, New York, 1968.

[18] A.E. Fitzgerald, C. Kingsley Jr., and S.D. Umans, *Electric Machinery*, McGraw-Hill Science/Engineering/Math, 6th ed., 2002.

[19] J. Gärtner, On large deviations from the invariant measure, *Theory Probab. Appl.*, 22 (1977), 24–39.

[20] R. Giglioli, P. Pelacchi, M. Raugi, and G. Zini, A state of charge observer for lead-acid batteries, *Energia Elettrica* 65 (1988), 27–33.

[21] W.B. Gu and C.Y. Wang, Thermal electrochemical modeling of battery systems, *J. Electrochem. Soc.*, 147 (2000), 2910–2922.

[22] B.S. Guru and H.R. Hiziroglu, *Electric Machinery and Transformers*, Oxford University Press, 2001.

[23] S. Haykin, *Unsupervised Adaptive Filtering: Volume I Blind Source Separation*, John Wiley Sons, Inc., USA, 2000.

[24] V. Johnson, A. Pesaran, and T. Sack, Temperature-dependent battery models for high-power lithium-ion batteries, *Proceedings of the 17th Electric Vehicle Symposium*, Montreal, Canada, October 2000.

[25] V. Johnson, M. Zolot, and A. Pesaran, Development and validation of a temperature-dependent resistance/capacitance battery model for ADVISOR, *Proceedings of the 18th Electric Vehicle Symposium*, Berlin, Germany, October 2001.

[26] I.-S. Kim, A technique for estimating the state of health of lithium batteries through a dual-sliding-mode observer, *IEEE Trans. Power Elec.*, 25 (2010), 1013–1022.

[27] A.D.M. Kester and W.C.M. Kallenberg, Large deviations of estimators, *Ann. Statist.*, 14 (1986) 648–664.

[28] P.R. Kumar and P. Varaiya, *Stochastic Systems: Estimation, Identification and Adaptive Control*, Prentice-Hall, Englewood Cliffs, NJ, 1986.

[29] H.J. Kushner, Asymptotic behavior of stochastic approximation and large deviation, *IEEE Trans. Automat. Control*, 29 (1984), 984–990.

[30] H.J. Kushner and H. Huang, Rates of convergence for stochastic approximation type algorithms, *SIAM J. Control Optim.*, 17 (1979) 607–617.

[31] H.J. Kushner and G. Yin, *Stochastic Approximation Algorithms and Applications*, 2nd ed., Springer-Verlag, New York, 2003.

[32] Z.Y. Liu and C.R. Lu, *Limit Theory for Mixing Dependent Random Variables*, Science Press, Kluwer Academic, New York, Dordrecht, 1996.

[33] L. Liu, L.Y. Wang, Z. Chen, C. Wang, F. Lin, and H. Wang, Integrated system identification and state-of-charge estimation of battery systems, *IEEE Transactions on Energy Conversion*, (2013), 1–12.

[34] D. Linden and T. Reddy, *Handbook of Batteries*, 3rd ed., McGraw Hill, 2001.

[35] L. Ljung, *System Identification: Theory for the User*, Prentice-Hall, Englewood Cliffs, NJ, 1987.

[36] L. Ljung, H. Hjalmarsson, and H. Ohlasson, Four encounters with systems identification, *Euro. J. Control*, 17 (2011) 449–471.

[37] L. Ljung and T. Söderström, *Theory and Practice of Recursive Identification*, MIT Press, Cambridge, MA, 1983.

[38] M. Milanese and A. Vicino, Optimal estimation theory for dynamic systems with set membership uncertainty: an overview, *Automatica*, 27 (1991), 997–1009.

[39] Online: http://www.mathworks.com/help/toolbox/physmod/powersys/ref/battery.html

[40] K. Ogata, Discrete-Time Control Systems, Prentice-Hall, Inc. Upper Saddle River, NJ, 1987.

[41] B.T. Polyak, New method of stochastic approximation type, *Automation Remote Control* 7 (1991), 937–946.

[42] G. Plett, Extended Kalman filtering for battery management systems of LiPB-based HEV battery packs. Part 1. Background, *J. Power Sour.*, 134 (2004), 252–261.

[43] M.A. Roscher, J. Assfalg, and O.S. Bohlen, Detection of utilizable capacity deterioration in battery systems, *IEEE Trans. Vehicular Tech.*, 60 (2011), 98–103.

[44] S. Rodrigues, N. Munichandraiah, and A. Shukla, A review of state-of-charge indication of batteries by means of AC impedance measurements, *J. Power Sour.*, 87 (2000), 12–20.

[45] D. Ruppert, Stochastic approximation, in *Handbook in Sequential Analysis*, B.K. Ghosh and P.K. Sen, eds., Marcel Dekker, New York, 1991, 503–529.

[46] M. Sitterly, L.Y. Wang, G. Yin, and C. Wang, Enhanced identification of battery models for real-time battery management, *IEEE Transactions on Sustainable Energy*, 2 (2011), 300–308.

[47] A.M. Sharaf, E. Elbakush, and I.H. Atlas, A predictive dynamic controller for PMDC motor drives, *Fifth International Conference on Industrial Automation*, Montréal, Quebec, Canada, June 11–13, 2007.

[48] T. Soderstom and P. Stoica, *System Identification*, Prentice-Hall, 1989.

[49] V. Solo, Robust identification and large deviations, in *Proc. 35th IEEE CDC*, Kobe Japan, Dec. 1996, 4202–4203.

[50] V. Solo, More on robust identification and large deviations, in *Proc. IFAC99*, Beijing, P.R. China, June, 1999, 451–455.

[51] V. Solo and X. Kong, *Adaptive Signal Processing Algorithms*, Prentice-Hall, Englewood Cliffs, NJ, 1995.

[52] K. Takano, K. Nozaki, Y. Saito, A. Negishi, K. Kato, and Y. Yamaguchi, Simulation study of electrical dynamic characteristics of lithium-ion battery, *J. Power Sour.*, 90 (2000), 214–223.

[53] O. Tremblay and L.A. Dessaint, Experimental validation of a battery dynamic model for EV applications, *World Electric Vehicle Journal*, vol. 3, at 2009 AVERE, EVS24 Stavanger, Norway, May 13–16, 2009.

[54] V.I. Utkin and H.-C. Chang, Sliding mode control on electromechanical systems, *Mathmatical Problems in Engineering*, 8 (2002), 451–473.

[55] S.R. Venkatesh and M.A. Dahleh, Identification in the presence of classes of unmodelled dynamics and noise, *IEEE Trans. Automatic Control*, 42 (1997), 1620–1635.

[56] A. Widodo, M.-C. Shim, W. Caesarendra, and B.-S. Yang, Intelligent prognostics for battery health monitoring based on sample entropy, *Expert Systems Appl.*, 38 (2011) 11763–11769.

[57] L.Y. Wang and G. Yin, Persistent identification of systems with unmodeled dynamics and exogenous disturbances, *IEEE Trans. Automat. Control*, 45 (2000), 1246–1256.

[58] L.Y. Wang and G. Yin, Asymptotically efficient parameter estimation using quantized output observations, *Automatica*, 43 (2007), 1178–1191.

[59] L.Y. Wang and G. Yin, Quantized identification with dependent noise and Fisher information ratio of communication channels, *IEEE Trans. Automat. Control*, 53 (2010), 674–690.

[60] L.Y. Wang, G. Yin, and J.F. Zhang, Joint identification of plant rational models and noise distribution functions using binary-valued observations, *Automatica*, 42 (2006), 535–547.

[61] L.Y. Wang, G. Yin, J.F. Zhang, and Y.L. Zhao, Space and time complexities and sensor threshold selection in quantized identification, *Automatica*, 44 (2008), 3014–3024.

[62] L.Y. Wang, G. Yin, J.-F. Zhang, and Y.L. Zhao, *System Identification with Quantized Observations: Theory and Applications*, Birkhäuser, Boston, 2010.

[63] L.Y. Wang, J.-F. Zhang, and G. Yin, System identification using binary sensors, *IEEE Trans. Automat. Control*, 48 (2003), 1892–1907.

[64] C.Z. Wei, Multivariate adaptive stochastic approximation, *Ann. Statist.* 15 (1987), 1115–1130.

[65] B. Widrow, J. Glover, J. McCool, J. Kaunitz, C. Williams, R. Hearn, J. Zeidler, E. Dong and R. Goolin, Adaptive noise cancelling: principles and applications, *IEEE Proc.*, 63 (1975), 1692–1716.

[66] G. Yin, S. Kan, L.Y. Wang, and C.Z. Xu, Identification of systems with regime switching and unmodelled dynamics, *IEEE Trans. Automat. Control*, 54 (2009), 34–47.

[67] G. Yin, V. Krishnamurthy, and C. Ion, Regime switching stochastic approximation algorithms with application to adaptive discrete stochastic optimization, *SIAM J. Optim.*, 14 (2004), 1187–1215.

[68] G. Yin, L.Y. Wang, and S. Kan, Tracking and identification of regime-switching systems using binary sensors, *Automatica*, 45 (2009), 944–955.

[69] G. Yin and Q. Zhang, *Discrete-Time Markov Chains: Two-Time-Scale Methods and Applications*, Springer, New York, 2005.

[70] G. Yin and C. Zhu, *Hybrid Switching Diffusions: Properties and Applications*, Springer, New York, 2010.

[71] G. Zames, On the metric complexity of causal linear systems: ε-entropy and ε-dimension for continuous time, *IEEE Trans. Automatic Control*, 24 (1979), 222–230.

[72] H. Zheng, H. Wang, L.Y. Wang, and G. Yin, Time-shared channel identification for adaptive noise cancellation in breath sound extraction, *J. Control Theory Appl.*, 2 (2004), 209–221.

[73] H. Zheng, H. Wang, L.Y. Wang, and G. Yin, Cyclic system reconfiguration and time-split signal separation with applications to lung sound pattern analysis, *IEEE Trans. Signal Processing*, 55 (2007), 2897–2913.

[74] X.Y. Zhou and G. Yin, Markowitz mean-variance portfolio selection with regime switching: a continuous-time model, *SIAM J. Control Optim.*, 42 (2003), 1466–1482.

[75] C. Zhu and G. Yin, Asymptotic properties of hybrid diffusion systems, *SIAM J. Control Optim.*, 46 (2007), 1155–1179.

Index

Q. He et al., *System Identification Using Regular and Quantized Observations*,
SpringerBriefs in Mathematics, DOI 10.1007/978-1-4614-6292-7,

Printed by Printforce, the Netherlands